Visual Analytics for Dashboards

A Step-by-Step Guide to Principles and Practical Techniques

Arshad Khan

Apress®

Visual Analytics for Dashboards: A Step-by-Step Guide to Principles and Practical Techniques

Arshad Khan
Tracy, CA, USA

ISBN-13 (pbk): 979-8-8688-0118-1 ISBN-13 (electronic): 979-8-8688-0119-8
https://doi.org/10.1007/979-8-8688-0119-8

Managing Director, Apress Media LLC: Welmoed Spahr
Acquisitions Editor: Susan McDermott
Development Editor: Laura Berendson
Coordinating Editor: Jessica Vakili

Cover designed by eStudioCalamar

Cover image designed by Freepik (www.freepik.com)

Distributed to the book trade worldwide by Apress Media, LLC, 1 New York Plaza, New York, NY 10004, U.S.A. Phone 1-800-SPRINGER, fax (201) 348-4505, e-mail orders-ny@springer-sbm.com, or visit www.springeronline.com. Apress Media, LLC is a California LLC and the sole member (owner) is Springer Science + Business Media Finance Inc (SSBM Finance Inc). SSBM Finance Inc is a **Delaware** corporation.

For information on translations, please e-mail booktranslations@springernature.com; for reprint, paperback, or audio rights, please e-mail bookpermissions@springernature.com.

Apress titles may be purchased in bulk for academic, corporate, or promotional use. eBook versions and licenses are also available for most titles. For more information, reference our Print and eBook Bulk Sales web page at http://www.apress.com/bulk-sales.

Any source code or other supplementary material referenced by the author in this book is available to readers on GitHub (https://github.com/Apress). For more detailed information, please visit https://www.apress.com/gp/services/source-code.

If disposing of this product, please recycle the paper

Table of Contents

About the Author

Arshad Khan is a versatile business intelligence/analytics and training professional, with 35+ years of experience, who has led many Tableau and SAP projects. He has consulted for Accenture, PwC, Deloitte, Chevron, Bose Corporation, Daimler Chrysler, Genentech, PepsiCo, Hitachi America, and many other blue-chip companies.

Khan has authored 17 books including six on business intelligence. He has taught many data visualization, dashboards, data analysis, BI, and SAP courses at nine universities including Golden Gate University, Santa Clara University, and the University of California (Berkeley, Santa Cruz/Silicon Valley, and San Diego Extensions). Khan has a graduate degree in chemical engineering and an MBA.

About the Technical Reviewer

Massimo Nardone is a seasoned cyber, information, and operational technology (OT) security professional with 28 years of experience working with IBM, HP, and Cognizant, among others, with IT, OT, IoT, and IIoT security roles and responsibilities including CISO, BISO, IT/OT/IoT Security Architect, Security Assessor/Auditor, PCI QSA, and ICS/SCADA Expert. He is the founder of Massimo Security Services, a company providing IT-OT-IoT Security Consulting Services, and a member of ISACA, ISF, Nordic CISO Forum, and Android Global Forum and owns four international patents. He is the coauthor of five Apress IT books.

Preface

Welcome to the world of *Visual Analytics for Dashboards*, where data meets design and insights come alive. In this rapidly evolving era of information overload, the ability to transform complex data into meaningful visual representations has become an indispensable skill for individuals and organizations alike. In this book, we embark on a journey to explore the art and science of visual analytics, with a specific focus on creating compelling dashboards.

The power of information lies not just in the abundance of data but in your ability to extract valuable insight from it. Visual analytics serves as your bridge between raw data and actionable knowledge, enabling you to gain a deeper understanding of trends, patterns, and relationships. Dashboards, in turn, act as a canvas where these insights are presented, facilitating informed decision-making, driving business growth, and fueling innovation.

As the demand for effective data communication grows, so does the need for skilled individuals who can transform complex information into visually appealing and intuitive dashboards. *Visual Analytics for Dashboards* is designed to cater to a wide range of readers, from beginners looking to grasp the fundamentals of visual analytics to seasoned professionals seeking to enhance their dashboard design skills. Whether you are a data analyst, business intelligence professional, data scientist, or simply someone interested in data visualization, this book aims to equip you with the knowledge and tools necessary to create impactful dashboards.

In *Visual Analytics for Dashboards*, we will delve into the key principles, best practices, and practical techniques for designing and implementing visually compelling dashboards. We will explore the various stages of the dashboard development process, from understanding user needs and defining goals to selecting appropriate visual encodings, designing effective layouts, and employing interactive elements.

It is important to note that while this book aims to provide comprehensive insights into the world of visual analytics for dashboards, the field is continuously evolving. New tools, techniques, and research emerge regularly, and it is our hope that this book will serve as a solid foundation for your journey into visual analytics, sparking your curiosity and encouraging you to explore further.

Our goal is to empower you with the knowledge and skills to create dashboards that not only present data effectively but also inspire action and drive positive outcomes. We believe that visual analytics is a powerful tool that can transform how we understand and interact with data, leading to informed decision-making and fostering a data-driven culture.

So, let us embark on this adventure together, as we dive into the exciting realm of *Visual Analytics for Dashboards*. May this book serve as a guide, unlocking the potential of your data and helping you create visually stunning and impactful dashboards that make a difference.

Happy exploring!

Arshad Khan

Introduction

Visual Analytics for Dashboards by Arshad Khan is a comprehensive guide exploring the multidimensional and complex world of business intelligence (BI). Khan takes his readers on a journey through the fundamental concepts, tools, technologies, and strategies that empower organizations to harness the power of data and make informed decisions.

One of the book's greatest strengths is in its ability to break down complex concepts into easily digestible explanations. Khan's writing style is clear, concise, and accessible, making it suitable for beginners and professionals seeking to deepen their understanding of data presentation and analysis. He balances technical details and practically implementable real-world examples, ensuring that readers grasp the practical applications of visual analytics in various areas.

Visual Analytics for Dashboards contains no fluff or repetition, and this makes this book concise yet complete. Khan has also included a bulleted list (read checklist) for almost every critical concept needed to implement a practical dashboard solution. Many readers will like to use this book because these bullet lists can be used as a reference.

The book also discusses the philosophy behind any good visual presentation with extensive visual examples. A brief but comprehensive discussion about different kinds of tools for data analysis, such as scorecards, key performance indicators, and their presentation to an end user on dashboards using different charts and tables, is invaluable.

Visual Analytics for Dashboards is a valuable resource for anyone seeking a solid foundation in data presentation and analysis. Khan's expertise and passion for the subject shine through in his writing, making the book engaging and informative. Whether you are a business professional, data analyst, novice, or an aspiring entrepreneur, this book provides the tools and knowledge to leverage data effectively and gain a competitive edge in today's data-driven world.

Ahmad Raza
Architect: Lantern Analytics Platform
Ex-CEO Modemetric (an analytics company)

It goes without saying that today's business relies on the most accurate, real-time data available to fuel the best decisions and possible outcomes. In any given enterprise, data professionals see data volumes growing at an average rate of 63% per month. Therefore, the challenge lies in contextualizing the business-critical data from the vast amounts of sources and formats generated every second. Advanced analytics and business intelligence have become an integral component of forward-looking, data-driven organizations seeking to maximize the value of their resources and convert their data into actionable insight.

The ability to visualize this analysis in the form of dashboards drastically streamlines and simplifies the decision-making process by presenting key metrics and the highest-value data for immediate consideration. Whether you are a student seeking to add dashboarding to your skillset or an analytics professional with line-of-business stakeholders, *Visual Analytics for Dashboards* provides you with the background, basics, and usable, easy-to-follow, real-life examples to put you in the driver's seat of your most valuable data so that you can harness it for easy and effective analysis. Arshad Khan takes us through the fundamentals and best practices of visual analytics, focusing on the most pervasive and highest-impact examples, as well as identifying the pitfalls to avoid—definitely the "go to" guide for anyone interested in optimizing their dashboarding technology and skills.

Jaime D'Anna

Head of Tech Partner Marketing, Neo4j

This remarkable book provides an extensive blueprint for crafting impactful dashboards. The author artfully outlines a comprehensive step-by-step methodology for constructing every component of a successful dashboard. A must read for dashboard designers across industries, irrespective of their expertise.

Naeem Hashmi

Strategic Advisor, Digital Health Solutions, Boston Scientific

Visual Analytics for Dashboards is an invaluable, user-friendly guide that imparts fundamental principles and specialized concepts for harnessing the full potential of dashboards in business intelligence. This comprehensive resource presents clear and accessible knowledge, empowering users to leverage this powerful tool effectively and enhance their analytics capabilities.

Dr. Miriam O'Callaghan
Associate Dean of Research and Scholarship
William Woods University

This book is exactly what corporate and education need! Insightful, designed to empower individuals with the skills to present intricate metrics in easily understandable data dashboards.

Dragana Celic
Industrial Engineer, University of California, San Diego Division of Extended Studies

CHAPTER 1

Dashboards

Business Intelligence

Objective and Technologies

Business intelligence, or BI, is a decision support system. It aims to help make business decisions, strategic as well as operational/tactical. It uses an assortment of resources and techniques for gathering, transforming, storing, and analyzing data, including processes, technologies, applications, quality, skills, practices, and risks.

The technologies utilized include data warehousing, multi-dimensional analysis (Online Analytical Processing—OLAP), data mining, analytical and statistical tools, querying and reporting tools, data visualization, dashboards, scorecards, etc. Together, various BI technologies enable multiple tasks relevant to data and information to be performed, including:

- Collection
- Integration
- Analysis
- Interpretation
- Presentation

Business intelligence is viewed differently by the two groups that use it—business and IT. While IT usually regards it as a tool, business views it as information.

A. Khan, *Visual Analytics for Dashboards*, https://doi.org/10.1007/979-8-8688-0119-8_1

Defining Business Intelligence

There is no single universally used definition of business intelligence. In 1989, Howard Dresner defined "business intelligence" as encompassing "concepts and methods to improve business decision-making by using fact-based support systems." In 1996, Gartner defined it as follows:

> *By 2000, Information Democracy will emerge in forward-thinking enterprises, with Business Intelligence information and applications available broadly to employees, consultants, customers, suppliers, and the public. The key to thriving in a competitive marketplace is staying ahead of the competition. Making sound business decisions based on accurate and current information takes more than intuition. Data analysis, reporting, and query tools can help business users wade through a sea of data to synthesize valuable information from it – today, these tools collectively fall into a category called "Business Intelligence."*

Business Intelligence Tools

BI tools are software applications used to retrieve, analyze, and present data. They have traditionally worked on stored data. They include a number of components that provide users with capabilities for analysis and visualization. Tool categories include:

- Reporting and querying
- OLAP
- Advanced visualization (dashboards and scorecards)
- Advanced analytics
- Data warehousing
- Data mining
- Business performance management
- Spreadsheets

Convergence of Disciplines

Performance management and business intelligence are two disciplines that need each other, as they meet the specific needs of organizations. Independently, they have struggled to provide business value and find acceptance within organizations. This shortcoming has been met through the consolidation of data, tools, and technologies under a unified platform, dashboards and scorecards, which represent their convergence. Together, these two disciplines offer a combination where the value is greater than the sum of the individual components.

Background

Business Dashboard Predecessor

The history of decision support systems can be traced back to the 1960s. In the 1980s, they were represented by the Executive Information Systems (EIS) systems. They served a useful purpose but had limitations that led to the development of the modern business dashboard.

Change in User Needs

Over the past couple of decades, user needs and requirements have changed dramatically. However, the basic requirement, that is, the ability to analyze performance, has not changed over time. Users have become far more demanding due to technological changes and the availability of powerful software tools combined with superior infrastructure and processing.

User expectations, all of which dashboards have been able to meet, include:

- Real-time and multi-dimensional analysis
- Ability to work with large data volumes
- Flexibility
- Ability to make changes on the fly
- Easy online access
- Drill-down, as well as the ability to slice and dice

Major Changes Since the 1990s

Major technological innovations in the 1990s encompassed both hardware and software. Data proliferated, and consequently, storage requirements also increased. Since the cost of storage and processing decreased simultaneously, they did not become a constraint. The development of enabling technologies, which included data warehousing and tools supporting reporting and analytics, enabled dashboards to leverage them extensively.

Data consolidation also became an important driver for dashboard use. Many organizations realized that they could use dashboards, an effective tool for managing enterprise metrics, to consolidate and integrate information from disparate sources and present it to its users.

More recently, in the past couple of decades, additional drivers have come into play. These include the emergence of cloud BI platforms, data and BI as service data products, as well as the ability to derive insight without worrying about the tools. Though such platforms simplify enterprise scale BI services, they fall short in providing the framework for presenting and consuming insight, which can help business users understand the derived insight and context, and also enable further exploration through drill-down into the details. Since dashboards help overcome this limitation, they have become very popular with business users, especially executives.

Dashboards

Car Dashboard

An automobile dashboard is a monitoring tool that monitors an auto's most important performance indicators, such as speed and fuel level. It displays the current state information at a glance. All indicators present current information, based on which the driver can take appropriate action if needed. The objective is to monitor performance:

- Ensure that it is on target.
- Identify any deviations or problems.
- Analyze the causes of issues.
- Take corrective action without delay.

Business Dashboard

A business dashboard is a monitoring tool derived from the car dashboard, which enables an organization's health to be monitored quickly. It visually displays the most important information required for monitoring on a single screen.

A dashboard provides a summarized view of the performance of an organization, business unit, department, group, or individuals. It is capable of quickly providing an overview as well as details. It enables an interactive and flexible environment, which can be easily updated when required by the business.

A dashboard provides valuable insight into large volumes of performance data through visualizations, summaries, trends, and deviations. It presents data using charts, gauges, traffic lights, etc. The tool's popularity can be attributed to its ease of use and the visibility that it provides to the business.

Dashboards are a powerful tool for business users due to the ease of use and the business insight that they provide. Their popularity can also be attributed to their ability to identify opportunities and threats, present results at both summary and detailed levels, and analyze historical data. Their level of adoption and engagement is significantly impacted by the user experience.

Definition

According to Stephen Few, a dashboard is a visual display of the most important information needed to achieve one or more objectives that have been consolidated on a single computer screen so that it can be monitored at a glance.

Objective

A dashboard aims to provide users with actionable information, which can be digested at a glance, in a format that is both intuitive and insightful. Dashboards leverage operational data primarily in the form of metrics and key performance indicators (KPIs). They are optimized so that users can quickly evaluate and react to the current situation and also communicate and share any issues and results with colleagues and partners.

Key Features

A well-designed dashboard fits on a single screen, eliminating the need to scroll for viewing the displayed information. It displays a limited number of KPIs and, typically, four to seven is considered an optimal range.

The following are the core characteristics and features of a dashboard:

- Enables performance information to be displayed on a single screen
- Provides a visual display
- Interactive and supports drill-down analysis
- Well organized from a display and layout perspective
- Condensed, primarily in the form of summaries; displays the most appropriate and relevant metrics
- Customizable
- Designed to communicate easily with its users
- Supports flexible analysis that static reporting tools cannot provide
- Provides links to other applications
- Presents data that is accurate and current
- Easy to access, understand, and use

Process Supported

There are four stages for monitoring information via a dashboard:

1. Update high-level situation awareness.
2. Identify and focus on particular items that need attention:
    - Update awareness of such an item in greater detail.
    - Determine whether action is required.

3. If action is required, access any additional information that may be needed to initiate an appropriate response.
4. Respond.

Dashboards must be designed to support this process.

Dashboard Types

Dashboard Classifications

Dashboards are implemented at various levels in organizations as well as business functions. The implementation can occur at different levels, such as enterprise, region, business unit, department, and group.

There are three dashboard classifications, which are defined by their role:

- Strategic
- Analytical
- Operational

Strategic Dashboard

The strategic dashboard focuses on an enterprise's high-level strategic performance objectives and KPIs. It provides a quick overview of the organization's health, including static data snapshots for various periods and trends. A strategic dashboard's objective is to enable decision-making by the executives. However, lower-level managers can also use the information provided on a strategic dashboard.

Analytical Dashboard

Analytical dashboards are analysis-focused, as they aim to gain insight from data collected over time—often the past month or quarter or year. They use such data to determine the following:

- What happened?
- Why did it happen?
- What changes should be implemented to improve future performance?

For example, by analyzing historical data, trend analysis can reveal why a particular product is underperforming in a specific region compared to other products.

Analytical dashboards use sophisticated models, such as what-if analysis and pivots, to identify patterns and opportunities, as well as align strategic goals with performance management initiatives. Such dashboards support more interaction with the underlying data through drill-down analysis. They are popular with business analysts, who are typically responsible for developing reports and providing meaningful analysis.

Operational Dashboard

An operational dashboard monitors business processes and performance metrics or KPIs. It monitors day-to-day activities, metrics, and KPIs, which change frequently and require immediate corrective action. The objective is to enable an organization to check, in real time, if its performance is on or off target and also by how much.

The KPIs an operational dashboard focuses on differ from those displayed on strategic or analytical dashboards.

Operational dashboards eliminate or reduce the need to distribute static reports. They are common in environments where it is required to act quickly on opportunities and issues pertaining to sales, marketing, help-desk, inventory management, etc. To make operational dashboards more valuable, they can be provided the ability to drill down to the underlying data source, generate alerts, and share the results in real time.

Dashboard Characteristics

Unique Design

The business drives dashboard requirements. The environments, needs, and modes of operation differ considerably from enterprise to enterprise. What works for one organization may not work for others. Even within an organization, the needs of diverse groups may be completely different and, hence, their dashboards will need to be designed differently.

Common Elements

Despite differences, most dashboards have some basic common elements. They:

- Provide useful and actionable data so that insight can be converted into action
- Are simple and easy to analyze
- Communicate easily and effectively
- Make effective use of data visualization
- Provide essential information without being accompanied by distracting information or graphics

Desired Dashboard Characteristics

The following are the basic characteristics of a well-designed dashboard:

- Provides the right level of information
- Provides effective visualizations
- Interacts with content and supports drill-down analysis
- Reliable and consistent
- Considered the single source of truth
- Intuitive, user-friendly visual user interface
- Supports the creation of personalized views
- Flexibility allows users to quickly change views, make updates, as well as add data, new content, and visualizations
- Provides easy access to users, with role-based access and delivery
- Provides information in the users' desired format, as well as through multiplc devices
- Accessible via a standard web browser as well as through various applications and devices, including mobile

- Integrated with other applications
- Features that enable significantly reduced IT support
- Uses accessibility techniques to accommodate end user's special learning requirements

High-Level Dashboard Product Features

Hundreds of software products are available to develop dashboards, ranging from simple applications to very sophisticated software packages. The key characteristics that they must support were described in the previous section. At a high level, the key features can be classified into the following categories:

- User experience
- Visualization
- Reporting
- Drill-down
- User management and authorization
- Collaboration

Common Dashboard Software Tool Features

Drilling

With drill-down capability, detailed data can be displayed and analyzed. For example, an enterprise sales report may be drilled down to determine regional sales. Subsequently, another drill-down can reveal the sales by product. Such drilling can be vertical or horizontal.

Filtering

Filtering limits the data displayed according to the specified screening criteria. For example, filters can be specified to display all products with a profit margin greater than 15% in California stores with weekly sales greater than $250,000.

Sorting

This feature enables sorts to be performed so that the best/worst performers can be easily identified by displaying them in ascending or descending alphabetical lists.

Pivoting

Pivoting enables the splitting of data into groups. For example, the annual sales report can be analyzed via pivot tables that provide quarterly figures by region and with summary year-to-date totals.

Calculating

This feature enables the development of new or unique calculations, such as average selling price (ASP).

Charting

This feature enables the display of data on the dashboard using various types of charts and graphs.

Each feature described above can be used independently or in combination with others. For example, a user may

- Drill down from the initial display
- Apply a filter
- Perform a calculation
- Sort the new result
- Display the results graphically
- Publish and distribute the results as a PDF via e-mail

Dashboard Benefits, Shortcomings, and Challenges

Benefits

Modern dashboards have been quite successful in meeting business performance management needs. They are capable of quickly providing an overview as well as details. Additional benefits associated with dashboards include:

- Enabling users to understand their business and monitor performance
- Providing information in a filtered, summarized, and easy-to-understand way
- Reducing the time required to analyze information, gain insight into data and trends, and take corrective action
- Increasing productivity; users can make faster decisions as they do not have to dig through irrelevant data
- Providing a single version of the truth
- Providing a consistent view of the business
- Empowering users and increasing motivation
- Enabling users who are not tech-savvy to customize their display, such as changing filters, subscribing, changing page layout, etc.
- Reducing costs and redundancy
- Delivering actionable information

Management can use dashboards to monitor performance and ensure that the strategy is implemented as planned. Depending on the requirements and the metrics being tracked, monitoring can be hourly, daily, weekly, monthly, or quarterly. When issues are identified, executives and managers can quickly take corrective action and ensure that progress toward the goals is maintained. If necessary, strategy can be changed to reflect the changed business conditions.

Shortcomings

Dashboards are often considered nice to have but not essential or critical. Despite their popularity, most dashboards live up to only a fraction of their potential. They often fail, not due to poor technology, at least not primarily, but because of poor design. Also, dashboards typically do not mean the same thing to different people in the organization, as there is a lack of understanding of its objective, what the tool aims to achieve, or its relationship to performance.

Many dashboards do not follow best practices for dashboards and data visualization, which is a critical element of any successful dashboard. They are more complex than needed and do not follow sound design principles. Also, some issues and limitations exist, especially challenges in handling data dependent on disparate data sources (which can be inconsistent and unreliable).

Low-cost dashboards are often developed quickly without identifying the objective. Impatient users with an urgent business need often drive their implementation, and their budget is limited. While such dashboards can meet some short-term needs, they are usually unable to scale or meet the organization's long-term needs. They frequently display the issues associated with ad hoc software development.

Challenges

The development and maintenance of dashboards is not easy. The variables and options that can be incorporated are numerous. Selecting KPIs from a large number of available indicators can be a challenging task. The challenge becomes even greater when the available visualization options for charts and other types of visuals are considered.

Having more options and choices does not automatically lead to the more effective use of visualizations. It often becomes a barrier to identifying the best way to design dashboards.

Obtaining Value from Dashboards

Determine the Objective

Before development, the type of dashboard that is required, as well as how it will be used, should be determined. The dashboard to be built, as well as its content, will depend on the overall objectives of the proposed dashboard and the business challenges it is trying to meet.

Determine the Expected Value

The real value of dashboards, which can be expensive to implement and maintain, comes from assessing the costs and benefits associated with each solution. Implementing a dashboard without first identifying the benefits is never a good idea. The expected benefits should be compared against the potential implementation and operating costs.

Value is derived from monitoring how the enterprise performs and how its goals and expectations are met, exceeded, or fall short. When organizations are provided previously unavailable insight into their performance, their benefits and profits increase. They can identify profitable products or services, improve sales, reduce reporting time, and cut costs efficiently. After estimating such benefits and associated costs, including implementation and operation, the expected value and ROI can be calculated.

Identify Users

Dashboards should meet the needs of a wide range of users, including executives, business users at different levels, and IT. Executives need to view, at a glance, the organization's performance. They also require the ability to drill down to lower-level data for further analysis and decision-making.

Business users typically need data at the operational level. On the other hand, IT needs a system that is easy to administer and maintain, is efficient from the IT resource utilization perspective, and enables content to be created that can be handed over to users for future enhancements. Therefore, it is imperative to identify users who will use the dashboard, as it must meet their needs.

Leverage Data Visualization

Most reporting tools are limited in requiring users to make a reasonable guess regarding where the answers to questions may reside. Data visualization is better at exposing hidden problems. When used on dashboards, it provides a distinct advantage over conventional reporting. Visualizations can provide information distilled from thousands or millions of data rows aggregated at many dimensional levels on a single page of graphics. Therefore, data visualization should be leveraged as much as possible.

Leverage Advantage over Conventional Reporting

Reporting is the foundation of business intelligence. It is designed for questions posed regularly for monitoring business performance. Reporting and analysis tools are used to identify problems that must be addressed. However, conventional reports are limited in their analysis. In many cases, they cannot identify problems at the summary levels (on which most reports are based). Detailed reports are inefficient as they do not support rapid analysis. Dashboards can help overcome these limitations.

Balance Conflicting Display Requirements

The information displayed on a dashboard must be carefully selected, useful, and actionable. If too little information is displayed, it will make the dashboard practically useless. However, if too much information is displayed, it can make it confusing, inefficient, cumbersome, and practically useless in many cases.

A dashboard should display only a few but important items so that it maximizes efficiency. It should contain more leading than lagging indicators. The numbers displayed should be complemented by context, such as highlighting the drop or rise in revenue through the use of explanatory text.

Dashboards should also highlight positive and negative performance deviations, change in KPIs from negative to positive or vice versa, trends, contributing factors, as well as alerts.

Selection and Implementation

Selecting a dashboard solution can be challenging as many factors need to be considered, including purpose, scope, features, functionality, integration, cost, etc. Additional factors to be considered include initial deployment cost, implementation period, and solution cost over time.

Cost Components

Cost is one of the most important selection variables, which is influenced by many factors. Costs that come into play include project implementation costs, training, value-added services, ongoing support, and consulting services (usually calculated as a percentage of the base cost).

The key variables that contribute to the total cost include:

- Vendor pricing structure
- License options
- User types (such as super users, developers, or data consumers)
- SaaS versus yearly or monthly subscriptions by user, group, or enterprise
- CPU or server license models, where the focus shifts from end users to data volumes and maintenance

Many cost-conscious managers tend to see expenses rather than returns when evaluating software applications. To select the best solution, they should consider the overall costs, including components identified previously, and the benefits. For more cost-conscious organizations, using trial versions and general free versions of solutions before making a selection is an alternative option.

Selecting the Dashboard Type

To choose the appropriate dashboard for any given business problem, questions must be asked regarding the objectives and goals. The responses to these questions will help identify the business metrics that need to be monitored as well as their measurement frequency. Based on the outcome of this exercise, the appropriate type of dashboard can be selected for implementation. It should be realized that in most cases, enterprises will require both analytical and operational dashboards in different parts of the organization—for long-term performance monitoring and day-to-day business decision-making.

Four basic questions, described in the following sections, can help determine which type of dashboard to use.

What Is the Objective?

A dashboard's goal must be aligned with the goals and corporate strategy before metrics are identified, developed, and implemented. If goals are strategic, an analytical dashboard is appropriate, while an operational dashboard is appropriate when constant monitoring is required.

Which Business Problem Needs to Be Solved?

This will identify the driver for the dashboard solution and the project scope. Depending on the answer, a strategic, analytical, or operational dashboard will need to be developed.

Who Will Use the Dashboard?

User roles will impact the dashboard features and design. For example, a strategic dashboard will meet the requirements of executives, while an operational dashboard will meet the needs of call center users.

What Are the Existing Performance Gaps?

Identifying gaps will enable metrics to be identified and developed, which can be used at different hierarchical levels. An operational or analytical dashboard might be appropriate depending on the performance being monitored.

Identifying the Technical Requirements

Three basic questions can help determine which type of dashboard to use:

- What is the data infrastructure status?

  If there exists a robust data warehouse-based infrastructure, leveraging available data becomes easy for any type of dashboard.
- What are the data latency requirements?

  If frequent data updates are required, and data requirements are time-dependent or real time, an operational dashboard is a logical choice.
- Where is the source data located?

  Analytical dashboards generally require more data types and, hence, more data sources; operational dashboards are relatively narrower in focus and typically leverage a single data source.

Implementation

As part of the solution evaluation, reviewing and analyzing other implementations via a study of white papers, site visits, etc., is a good idea. Study cost figures provided by vendors for their products, which can include components costs, solution cost, and average deployment cost based on the number of users or various licensing fees/structures. Such numbers can provide additional insight into what to expect when implementing the dashboard solution.

The implementation period for dashboard projects can vary significantly. Most organizations want to implement it quickly once they have made a decision to proceed. The time to implement a solution can be quite different based on two scenarios:

- The company has a robust BI infrastructure in place.
- The company is considering a dashboard for the first time.

Summary

This chapter delved into the realm of dashboards within the context of business intelligence. It elucidated BI's role as a decision support system, employing various technologies for data gathering, analysis, and presentation to aid in strategic and operational decision-making. The chapter elucidated the evolution and significance of dashboards, tracing their lineage from EIS systems to modern-day analytical and operational dashboards. Additionally, it delineated the fundamental characteristics, types, benefits, shortcomings, challenges, and implementation considerations associated with dashboards, offering a comprehensive overview of their utility and impact in contemporary business environments.

CHAPTER 2

Scorecards

Understanding Scorecards

What Is a Scorecard

A scorecard is one level above a dashboard in the business decision-making process. It is primarily used to align operational execution with business strategy. A scorecard is a decision-making tool that displays the performance of an organization at a point in time. It provides periodic snapshots of performance associated with the organization's strategic objectives and plans.

A scorecard's objective is to focus the business on strategic objectives by monitoring execution and, subsequently, mapping the results back to a specific strategic objective. It aligns strategic goals and enables implementation of the strategy through monitoring and measurement against targets. While a scorecard basically measures against goals, a dashboard does not typically compare against goals.

A scorecard is a tabular display of KPIs, which reflects different perspectives and the organization's strategic objectives, along with their respective targets. It shows how each measure is performing against its specified target. By monitoring key indicators, management can ensure performance targets are met and the company's strategic goals are achieved.

Comparing Dashboards and Scorecards

Dashboards and scorecards can contain similar information. The key difference is that a scorecard displays information at the highest strategic level of decision-making. Compared to a dashboard, a scorecard's view is more static and just provides a performance snapshot. The terms dashboard and scorecard are frequently used interchangeably. However, they differ in how they are used and how they achieve their objectives. While scorecards measure against goals, dashboards need not do so.

A. Khan, *Visual Analytics for Dashboards*, https://doi.org/10.1007/979-8-8688-0119-8_2

Dashboards and scorecards are sophisticated information systems that are based on a BI foundation and data integration infrastructure. They are used by businesses to help them achieve their strategic objectives through measurement, monitoring, and management of business performance. These tools can measure the past, monitor the present, provide forecasts, and enable organizations to adjust their strategy in real time, as needed, so that performance can be optimized.

Scorecards ensure that employees are accountable for their performance. They highlight high-priority objectives and assign owners to each KPI. On a scorecard, metrics are organized and displayed by status, which enables problem areas to be quickly identified and corrective action to be taken in a timely manner.

Implementing and Using a Scorecard

Implementation Issues

There are five key issues that need to be considered when building a scorecard. The issues are:

- Metrics
- Format
- Standards
- Collection
- Use

Metrics Issue

A scorecard's effectiveness depends on the type and quality of the information it tracks. A common mistake is to measure too many items and lose focus on the key metrics that really matter. Another mistake is to collect the most easily accessible data. However, only the right data must be collected, as users need to focus only on the data that helps decision-making. For example, metrics for managing product costs will be different compared to those used for improving quality. Therefore, it is to be expected that product and quality managers will need to monitor different measures and KPIs.

Another issue is collecting too much data, which causes loading and information overload. Data should not be collected for the sake of data. Only actionable data should be collected and presented.

Format Issue

Scorecards are developed in many formats and applications. These include simple spreadsheets, tables in Word documents, sophisticated software embedded in a company's ERP system, project management applications, etc.

Standards Issue

A common mistake of novice scorecard designers is to use standard measures. However, businesses operate uniquely and, therefore, what may be right for one organization or division may not be appropriate for another. When building a scorecard, it should be tailored to deliver the specific information that is needed by its users.

Collection Issue

Many methods can be used to collect data required for creating a scorecard. Some of the data can be collected easily and quickly. However, some data might not be easy to collect and, hence, the tendency is to avoid such data. The most relevant and important data should be collected, even if it is difficult to collect, store, and analyze.

Use Issue

The objective of collecting information for a scorecard is to use it—not just display it for informational purposes. Just displaying a lot of metrics and KPIs is not beneficial if that data is not used for decision-making. Keep in mind that what is done with the results will determine if the scorecard will be successful or not.

Making an Effective Scorecard

An effective scorecard has the following characteristics:

- Focused
- Comprehensive

- Proactive
- Analytical
- Actionable

To make a scorecard focused, it should be aligned with the strategic business plan and driven by objectives and targets. It should avoid too many graphs, tables, and interpretations, as affluence leads to confusion or information glut, leading to rejection. Instead, it should focus on exceptions and place more effort and space on reporting, highlighting, and root-cause analysis of items that are performing too well or too poorly.

The scorecard should be made comprehensive without displaying extraneous information not required for decision-making. High-level information in an enterprise-level scorecard should be linked with lower-level scorecards. For example, if the enterprise scorecard is displayed on the home page, drill-down to lower-level scorecards, such as an operations scorecard, should be enabled.

To make a scorecard proactive and analytical, it should support analytics that answers the question, "Why is it happening?" It should support analysis and forecasting by using associated tools like data warehouses/OLAP, as well as leading and lagging indicators. It should also provide trend analysis and forecasts. A scorecard should contain information that makes it actionable. It should report exceptions as well as their reasons. After reasons are provided, the next question will be "What are we going to do about it?"

Balanced Scorecard

Background

A balanced scorecard is a strategic performance management tool used by management to monitor the organization's performance compared to its strategic goals. It helps align business activities with the organization's strategic goals. The concept was developed because traditional performance measurement systems were typically based only on financial performance indicators and were therefore inadequate for managing the organization's performance.

Balanced scorecards are widely used by all types of organizations, including large and small companies, non-profits, and governments.

Objective

The objective of a balanced scorecard is to provide a balanced and integrated view of an organization's performance. It aims to be an essential decision-making tool that aligns everyone around a common strategy and goals. A balanced scorecard aims to provide an enterprise view of an organization's overall performance. It integrates financial measures with other KPIs, using conventional financial metrics as well as strategic non-financial metrics.

What a Balanced Scorecard Does

A balanced scorecard implements and manages the enterprise strategy at all levels. It links objectives, initiatives, and measures to the organization's strategy. A balanced scorecard converts a mission statement into a comprehensive set of objectives and performance measures that can be quantified and appraised. It enables managers to monitor how the business is being executed and also the impact of corrective actions. It also improves communication internally as well as externally.

Four Balanced Scorecard Perspectives Strategic Areas

A balanced scorecard has four perspectives, which cover the main strategic focus areas of a company, as shown in Table 2-1. These are:

- Financial
- Customer
- Internal processes
- Learning and growth

Table 2-1 provides examples of objectives and goals for the four perspectives.

Table 2-1. Balanced scorecard objectives and goals

Perspective	Objectives	Goals
Financial	Increase revenue	Increase net revenue by 5%
Customer	Achieve a high customer rating	Increase the customer satisfaction rating to 95%
Internal business processes	Increase operational efficiency	Reduce product development cycle time
Learning and growth	Have a skilled workforce	Provide cybersecurity training to 100% of the workforce

A balanced scorecard is used as a template for designing strategic objectives, measures, targets, and initiatives within each of the four perspectives. Individual organizations tailor the four perspectives to suit their own needs. However, the basics have not changed much over time.

Financial Perspective

The financial perspective pertains to an organization's financial objectives, which shows how it is viewed by its shareholders from a growth, profitability, and risk perspective. It encourages the identification of a few relevant high-level financial measures, which help choose measures that answer the question, "How do we look to shareholders?"

Examples of the financial perspective include revenues, earnings, return on capital, cash flow, return on investment, project profitability, and backlog.

Customer Perspective

The customer perspective shows how the organization is viewed by its customers. It deals with the strategy for creating value and differentiation. It covers customer measures such as customer satisfaction, market share goals, as well as product and service attributes. This perspective encourages the identification of measures that help answer the question, "How do customers see us?"

Examples of the customer perspective include market share, customer satisfaction index, customer loyalty, and customer ranking.

Internal Processes Perspective

The internal processes perspective shows how well an organization manages its operational processes. It deals with the strategic priorities for various business processes, which create customer and shareholder satisfaction. It covers internal operational goals and outlines the key processes necessary to deliver customer objectives. This perspective encourages the identification of measures that answer the question, "What must we excel at?"

Examples of the internal processes perspective include productivity rates, quality measures, timeliness, safety incident index, and project performance index.

Learning and Growth Perspective

The learning and growth perspective focuses on the organization's ability to continue improving and creating value. It reviews how the organization learns and grows. It focuses on the priorities for creating a climate that supports organizational change, innovation, and growth. This perspective also covers intangible drivers of future success, such as human capital, organizational capital, as well as skills, training, organizational culture, and leadership. This perspective identifies measures that answer the question, "How can we continue to improve and create value?"

Connection Between Perspectives

Strategic planning, the balanced scorecard way, means implementing change from different aspects of a business, which are the various perspectives. The basis for the balanced scorecard is that the perspectives are interconnected and, therefore, any changes in one will have ramifications for the others.

The balanced scorecard process starts at the bottom with the learning and growth perspective. If the skills, culture, leaders, and management information are aligned with the organization's strategy, it will create effective and efficient business processes (internal processes perspective). Effective and efficient product delivery, customer relationships, innovation, and regulatory processes will ensure that the organization's offerings meet the customers' needs.

The customer perspective shows various components of the organization's offerings, such as products, service relationships, and brand. The combination of satisfied customers and efficient business processes produces growth, lowers costs, and enables better use of the organization's capital, which results in increased profits and shareholder value.

Comparing Scorecards and Dashboards

There are three common characteristics of dashboards and scorecards. Each has three applications, three layers, and three types.

Three Common Characteristics

Applications

Dashboards and scorecards are characterized by three applications. The three common tightly integrated applications, which can be a software program or a group of programs, are

- Monitoring application:

  Conveys information at a glance, which includes visual elements, charts, graphs, alerts, and symbols

- Analysis and reporting application:
  Analyzes exception conditions, which includes analytics, forecasting, visual analysis, and reporting

- Management application:

  Improves coordination and collaboration, which includes annotations, strategy maps, and workflows

These applications, which may or may not be distinct programs, provide specific functionality. For dashboards and scorecards, they help display information quickly and efficiently, highlight exceptions, and improve execution through easy communication and collaboration throughout the organization.

Layers

Dashboards and scorecards are characterized by three common information layers:

- Monitoring:

  Displays data, such as charts that monitor key performance indicators

- Analysis:

  Consists of summarized data for determining the root cause of identified problems

- Action:

  Displays detailed underlying data, which helps determine the corrective action to be taken to solve an identified problem

Starting at the top, users can drill-down through successive layers so that the root cause of a problem can be identified and, subsequently, take appropriate corrective action.

Types

The three common types are as follows:

- Strategic:

  It is geared more toward management rather than analysis and monitoring.

- Tactical:

  It is geared more toward analysis rather than monitoring.

- Operational:

  It is at the lowest level and tracks business processes, with the primary objective being monitoring rather than analysis or management.

Differences

Dashboards and scorecards have distinct differences because they focus on different levels of the organization. A dashboard is a performance monitoring system with an operational focus that measures performance. On the other hand, a scorecard is a performance management system that focuses on the progress toward achieving strategic objectives. Scorecards generate long-term considerations and actions, while dashboards display results that usually require immediate attention.

A dashboard measures performance, while a scorecard charts progress. A dashboard's primary users are managers and lower-level employees, while scorecards are primarily used by executives and managers. For dashboards, the information update frequency varies from real time to daily, compared to scorecards, which provide periodic snapshots. The top-level display for dashboards is charts and tables, while for scorecards it is symbols and icons for highlighting the status of KPIs and trends.

Dashboards monitor performance for a shorter time horizon. Depending on the organization, such monitoring can cover daily, weekly, hourly, or even real-time periods. On the other hand, scorecards look at longer time spans, which can cover quarterly, monthly, or even weekly periods. While dashboards are operational and occasionally tactical in nature, scorecards are primarily strategic and tactical.

	Dashboard	Scorecard
What it measures	Performance	Progress
Types of users	Managers and lower-level employees	Executives and managers
Update frequency	Varies from real-time to daily	Provide snapshots
Top level display	Charts and tables	Symbols and icons for highlighting KPIs and trends
Monitoring time horizon	Short period	Longer period
Type	Primarily operational and occassionally tactical	Primarily strategic but can also be tactical

Ensuring Effectiveness

There needs to be an integration of objectives, programs, and KPIs. Simply displaying KPIs on a dashboard or scorecard will not improve performance. Besides measuring and monitoring KPIs, there needs to be active management and appropriate actions, when needed, so that targeted goals are realized. Execution should be monitored and the results mapped back to the strategy.

Summary

This chapter explored the role of scorecards in aligning operational execution with strategic goals, distinguishing them from dashboards by their focus on goal measurement and performance snapshots. It covered key implementation considerations and how to make an effective scorecard. It introduced the balanced scorecard framework, emphasizing its four perspectives—financial, customer, internal processes, and learning and growth—and their role in providing a comprehensive view of organizational performance. Additionally, it compared scorecards and dashboards, discussing their common characteristics and differences.

CHAPTER 3

Key Performance Indicators

Measuring Performance

Why Performance Is Measured

Organizations measure performance for three main reasons:

- Learn and improve performance.
- Provide external reporting for shareholders and compliance requirements.
- Monitor and control internally within the organization.

A widely held view is that only what gets measured gets done and, hence, unless something is measured, it cannot be managed and controlled. Therefore, every enterprise collects performance data, which it uses for monitoring and improving performance. For this purpose, performance management analytics tools and techniques are used so that collected data can be analyzed and used to improve performance.

Measurement Requirements

There exist many measurement drivers, with the first being measuring for learning and improving performance. Key performance indicators are used extensively, especially to learn and improve. They provide information to users for making informed and, hence, better decisions so that performance can be improved. Monitoring KPIs enables management to remain informed, assumptions to be challenged, and ongoing learning and improvement to be provided.

A. Khan, *Visual Analytics for Dashboards*, https://doi.org/10.1007/979-8-8688-0119-8_3

Another driver is external and compliance reporting requirements. KPIs are measured to ensure that these requirements can be provided easily and accurately. Such requirements are generated due to the need to inform external stakeholders and shareholders, compliance with external reporting requirements, and special requests for information or data. Some of this information can be requested periodically or on an ad hoc basis. Many reports and associated indicators, such as quarterly financial reports, need to be provided on a regular basis as they are mandatory.

Measurements for controlling and monitoring are another driver. KPIs are routinely used to monitor and control the performance and behavior of employees at all levels of the corporate hierarchy. They are used to set performance goals and monitored to ensure that they have been met. Whenever variance exists between a goal and its achievement, appropriate corrective action is initiated.

Linking Measures

Measurements aim to eliminate variances and improve conformity. To achieve this objective, measures are linked to financial and associated rewards. Such reward structures need to be implemented carefully. A poorly designed system can be self-defeating when it causes employees to focus on meeting the measure targets, not on performance improvement.

Metrics

Measures and Metrics

A measure is a unit or standard of measurement, which consists of a number and a unit. For example, speed is expressed in miles per hour (mph) and temperature in degrees (°F). To be useful, measures should be defined and used consistently across the organization, which enables easy sharing, aggregation, and comparison. In the business environment, a measure represents a piece of business data that is related to a dimension, such as revenue by month. In this case, revenue is the measure (dollars), while time is the dimension (month).

Measure and metric are basically the same. A metric is used in business performance and has a goal or performance associated with it. It is a measurement standard which provides a target value that must be achieved for success. For example, business metrics can be revenue per customer, gross margin, and average cost per customer.

Difference in Objectives

The objective of metrics and related targets is to monitor progress toward the achievement of strategic goals and, ultimately, the implementation of the organization's strategy. Metrics are, ultimately, the key to the success of any dashboard or scorecard.

For using the right metrics, differences between strategic and operational measures need to be understood, as they have different objectives. Operational measures monitor measures in a shorter time frame, which can be daily or even more frequently, including real time. Strategic measures monitor indicators that indicate long-term performance. They can typically cover monthly, quarterly, or annual periods.

Balancing Metrics

There are thousands of indicators that can be measured even in a small organization. Therefore, a challenge exists to identify and pick the appropriate metrics that should be measured. The key is to measure items that can help an organization to monitor and improve its performance. Metrics should be actionable and enable performance improvement. Those that are nice to know but do not provide any value or trigger actions should be avoided.

Selected metrics should enable the enterprise performance to be evaluated. They should include financial and non-financial measures, existing metrics and custom metrics, as well as leading and lagging indicators. Analysis effectiveness is enhanced when both qualitative and quantitative metrics are utilized. The tendency of most organizations is to only use quantitative metrics (or numbers), as they are available, easy to extract, and can be converted into meaningful metrics that the business can use for performance management. However, quantitative metrics should be balanced with qualitative, that is, non-numeric, metrics, as they can provide additional insight and highlight issues with customers and stakeholders.

How to Obtain the Best Results

Organizations that follow best practices know which metrics are necessary and provide the best results. They focus on them, rather than use indicators for controlling employees, which should be avoided. Indicators required for external reporting should be separated if they are not relevant internally. This is beneficial as it makes monitoring more focused and targeted, screens out data that is not required, and helps improve performance.

Key Performance Indicators

What Is a Key Performance Indicator?

A key performance indicator (KPI) is a type of measurement used to gauge or compare performance. It is a metric that is tied to a target. However, a metric is not necessarily a KPI. KPIs reflect critical success factors and organizational goals. They are evaluated over a specific time period and compared against acceptable values, historical performance, or targets.

KPIs help reduce uncertainty. They are used to determine how far, above, or below a metric is compared to a pre-determined target. They reduce a complex organization's performance to a few key indicators that can be easily monitored. Just like doctors use only a few key health indicators, like blood pressure and temperature, organizations also need to measure and monitor the most important and relevant KPIs. When KPIs are not easily available, proxy indicators can be used.

KPI Characteristics

The following are the key characteristics of KPIs:

- Relevant
- Based on accurate and valid data
- Quantifiable
- Reflect strategic goals of the organization
- Impact the organization
- Easily understood across the organization
- Based on corporate standards
- Do not change frequently
- Measured frequently
- Trigger action at appropriate levels

KPIs Variety

Organizations are structured uniquely and operate differently with different processes and environments and, therefore, their performance criteria are different. No universally accepted set of KPIs exists for any business or function because they depend on the organization's business strategy, processes, and priorities, which can vary significantly across different organizations. Consequently, the KPIs used and their relative importance will vary significantly across organizations.

The range of KPIs that can be used varies significantly, such as:

- Top product revenue
- Sales growth year-to-date
- Percentage of on-time deliveries
- Number of complaints received
- Number of defects for new products
- Percentage of late payments
- Return on equity
- Profit margin
- Customers lost or gained in the previous quarter

KPIs to Use

When the right metrics are used, they pinpoint the organization's strengths and areas where its performance is superior compared to the industry and competitors. They can also point out weaknesses and identify areas where performance is lagging compared to the competition or the industry in which it operates. Using limited and correct metrics screens out unnecessary data, presents only the data needed to make well-informed decisions, and helps managers avoid sifting through a lot of distracting information.

Dashboards need to be unique to the specific organizations implementing them. The content should be customized for the organization and only include metrics that will help in meeting its goals. Only metrics unique to the organization should be included. Performance metrics are unique to each organization, though there may be similarities with other organizations. Using existing repackaged measures is risky as they may not meet the organization's needs.

It is not unusual to find that over half of the required strategic indicators are unique. Therefore, they have to be designed from scratch. Such unique indicators tend to be among the most relevant and useful metrics.

Benefits and Issues

Metrics and KPIs are the building blocks of many dashboard visualizations. They help management and users in tracking how well they are performing compared to their goals and objectives. The selected KPIs, if correctly selected, will provide a true picture of how an organization is progressing toward achieving its goals.

A problem with KPIs is that they can be difficult to implement and use. They may have limitations regarding extracting data from some sources or availability of clean data sources or may be difficult to change without causing disruption. Also, inadequately defined KPIs targets can be unrealistic and distort the perception of performance.

Using and Managing KPIs

KPI Management

KPIs highlight an organization's performance in its quest to achieve its strategic goals. Their management involves defining what the organization needs to do to achieve its objective, such as growing revenues at a certain rate and then prioritizing them. The next step involves creating indicators that measure the progress toward the achievement of those goals. Finally, progress is analyzed to identify what works and any changes that need to be implemented to ensure that the objectives are achieved.

Metrics As Predictors of Success

If dashboards are well designed, they can enable organizations to track their strengths and weaknesses, monitor performance, and aid in planning. When dashboards are based on irrelevant metrics, they waste resources and do not help in making the organization profitable and successful. Therefore, metrics used on a dashboard should enable the organization to succeed in achieving its goals.

A dashboard can be a critical decision-making tool when it contains metrics that are the best predictors of success. When such metrics are used, they enable valuable insight that management can use and act upon. Therefore, metrics should be used that are aligned with the organization's goals—operational and/or strategic.

Dashboard KPI

Dashboard KPIs should not be selected randomly. Since space on a dashboard is limited, only the most important and actionable metrics that are success predictors should be used. Nice to have metrics that do not trigger any actions should be avoided. Many dashboards are based on irrelevant KPIs, which are selected randomly or without any evaluation or flawed measurement techniques and metrics. These mistakes should be avoided.

Complicated and artistic presentations consisting of elaborate graphics and animation should be avoided. A dashboard should be able to communicate its information to users easily and quickly without any distractions. Users should easily be able to understand the metrics that are displayed so that they can quickly analyze them and take appropriate action. It is also good practice to display the relevant time, such as the data refresh date and time, on a dashboard.

Number of Metrics to Include

Humans can only absorb limited information at any time. Therefore, to avoid cluttering a dashboard, which creates brain processing overload, it should only display a handful of KPIs. The recommended guideline is four to seven KPIs on a dashboard screen. If there are more metrics, which is reasonable if measuring an end-to-end business activity, there are options to handle such a scenario. These include creating dashboard tabs, developing hierarchies of metrics, and using folders, tables, and drill-downs, which will make the dashboard display clear, simple, and easy to navigate.

Ownership

Organizations need to hold users accountable for the outcome of the measures. If no one is accountable, metrics won't have any practical impact. Therefore, it is advisable to hold an individual accountable, even if a team manages the process or task being measured.

Metrics Review

All metrics have a natural life cycle and, over time, lose their impact as processes are streamlined to the point where additional gains are not worth the effort. Organizations also change over time as they attempt to improve or meet business challenges, which forces them to add new metrics. Therefore, to keep their dashboards current and useful, their metrics will need to reflect the current environment and usage.

To ensure the effectiveness of metrics and the performance management system as a whole, continuously monitor the usage of metrics. Dashboard teams should periodically prune underused metrics after consulting with the business. They should also monitor newly deployed metrics to quickly identify problems that users may be having with the existing metrics or views.

Data Quality Review

Users should have complete confidence in the data displayed on a dashboard. Therefore, it is imperative that the dashboard content be periodically reviewed and validated, which can be done through a regular auditing process that includes the following:

- Key indicators are reviewed and subsequently traced back from the results to the source data.
- Calculations and formulas are checked.
- Data quality is validated.
- Verification that the correct triggers are used to indicate stop lights (red, green, and yellow).

Developing KPIs

Identifying New KPIs

Due to the unique way in which businesses are run, it is not unusual for a need to exist to develop custom metrics. Traditional indicators are not suitable when users want to measure something that has not been previously measured such as a new process and, therefore, off-the-shelf or traditional indicators are not suitable in such a case.

If an organization has already identified short- and long-term goals, defining new metrics becomes easier. To start designing a KPI, determine the questions for which answers are sought. Then identify one or two key performance questions (KPQs) for each strategic objective, which will lead to the new KPIs that need to be developed. KPQs help capture more valuable information, guide discussions, and facilitate communication.

Creating New KPIs

New metrics can be easy or difficult to create, depending on the complexity. Usually, creating a metric is easy if it involves customizing an existing metric. Creating a new metric entirely from scratch can be laborious. Some areas are relatively easy to quantify, while others present a challenge in the selection and design of appropriate KPIs. For example, KPIs for sales or the number of defects can be easily designed as they are quantifiable. However, others, like customer satisfaction, are not so easy to measure and design.

The time to create a metric can vary quite a bit, depending on the requirement. Depending on the complexity, creating a new metric can take weeks or even months before an agreement is reached on what exactly should be measured and how it should be measured. Scorecards, which are more strategic in nature, tend to create more unique metrics in new areas of the business than dashboards, which tend to measure established processes.

Participants in KPI Design and Development

The functional and content requirements should be determined by the users, as they understand the business and how its performance can be analyzed. However, IT should be involved in the design and development as it provides valuable input regarding data sources to be used, validating and loading data, implementing roles and authorizations, setting up alerts and emails, as well as other technical tasks. During discussions with the business, IT can evaluate the existence and condition of the data required to calculate the proposed metrics. IT is also responsible for installing and maintaining the infrastructure required to support the dashboard, maintaining data integrity, and keeping the application operational.

Data Requirements

Data Collection Methods

Data collection is the technical aspect of the process. In this step, the data collection methods are evaluated to determine their strengths, weaknesses, and their appropriateness for the dashboard project being undertaken.

Many methods are used for collecting data including interviews, questionnaires, surveys, focus groups, workshops, as well as documentation reviews. The data collection task involves identifying and describing the method that will be used to collect data.

Identifying the Data Source

In this step, the data source is identified. An important associated task requires determining if the identified data can be accessed easily and cost-effectively. In this step, it is also determined if the data source can provide accurate and reliable data. If the analysis leads to a negative conclusion, then an alternative data source or collection method will need to be used.

Data Collection Responsibility

It is always a good idea to assign data collection responsibility. A single user or group can be assigned the responsibility for collecting and updating data. The responsibility for measuring can be assigned to an internal user or function. It can also be assigned to an external entity because many companies outsource their collection of specific indicators, such as customer satisfaction, reputation, brand awareness, and employee satisfaction.

Data Collection Frequency

The frequency of data collection varies and depends on the type of dashboard, strategic or operational, as well as the metrics being monitored. The timing and collection frequency depends on the business needs, as well as IT infrastructure limitations. The most common data load frequency is once per day.

Some indicators are collected continuously and in real time. The data collection frequency can be on a daily, monthly, quarterly, or annual basis. As an example, traffic to a website might need to be monitored continuously, while customer satisfaction might be limited to one to two times a year. The most common data load time is early in the morning.

If data is not collected at a rate required to support monitoring and analysis, it can limit performance assessment. For example, many organizations survey employees annually. However, such large gaps in data collection and assessment can be a serious issue. Such a time gap can prevent timely assessment and corrective actions.

Cost Components

There is cost and effort associated with creating and maintaining performance indicators, which are not recognized by many executives and managers. Costs components include measurement cost, data collection cost, administration cost, as well as performance reporting and analysis costs. These costs must be identified and considered when evaluating the overall cost.

Temporary Indicators

Some indicators are developed with the understanding that they will only be used for a specific period, such as the duration of a major project or initiative. An issue is that a significant number of temporary indicators are introduced but not discarded after completing their purpose. They continue to be collected indefinitely because no one identifies the need to stop collecting and using them. For such indicators, sunset or review dates should be specified when it can be determined if they are still needed. A process should be initiated for reviewing all indicators to ensure that they are still required and used.

Summary

This chapter delved into the significance of measuring performance within organizations, outlining three primary purposes: learning and improving performance, external reporting for stakeholders, and internal monitoring and control. It emphasized the critical role of KPIs in this process, which provide actionable insights for decision-making and performance improvement. The chapter discussed the diverse measurement drivers, from enhancing performance to meeting compliance requirements, and underscored the importance of selecting relevant metrics. Additionally, it explored the characteristics and variety of KPIs, offering guidance on their usage, management, and development, as well as data requirements.

CHAPTER 4

Dashboard Requirements

Objective and Users

Objective

Software projects are doomed to failure if they are not based on clear requirements, which must be defined before a project is started. Dashboard projects are not any different and, therefore, must be based on business and technical requirements that have been pre-approved.

Dashboard objectives can be strategic or operational. They can vary according to the types of users. For example, requirements will be quite different for a call center dashboard, a single department dashboard, or a dashboard for the enterprise senior executive team.

Users

From a business user's perspective, efficient use of a dashboard comes with a number of prerequisites and requirements. These need to be identified based on the user type—executive, manager, lower-level employees, etc. A dashboard will become an efficient tool for them only if the specified requirements are delivered.

To ensure dashboard success, from an implementation and usage perspective, the inclusion of key players and collaboration is imperative. If end-user needs are determined to start with, it lays the foundation for ultimate success. Since a dashboard's objective is to inform users and enable them to take action quickly, their metric needs must be understood and incorporated into the requirements. The specific user needs can be determined using a variety of techniques, such as interviews and workshops.

Many groups and roles are involved in the overall requirements-gathering process as well as in project implementation. The key groups involved in designing and developing a dashboard are:

A. Khan, *Visual Analytics for Dashboards*, https://doi.org/10.1007/979-8-8688-0119-8_4

- End users (includes executives, managers, analysts, and others)
- Business analysts
- IT team
- Project manager

Functional Requirements

High-Level User Requirements

Developing an effective dashboard application is a collaborative process. Stakeholders and users should be included in the process to determine the requirements that the application will be expected to deliver.

The main user requirements for designing a dashboard are:

- Content
- Accuracy
- Presentation
- Display
- User action
- Access
- Output
- Distribution

Basic Requirements

A dashboard should contain information that answers key business questions. It should enable a user to quickly digest the key metrics and identify any exceptions so that appropriate investigation and/or action can be taken. Users should be able to understand the cause and effect and, also, identify trends and correlations to key drivers. They should be able to obtain a clear picture in a single integrated view of the current status.

The data displayed on a dashboard should be:

- Accurate
- Clear
- Consistent
- Unambiguous

A dashboard should support users in analyzing the presented data through filtering, sorting, grouping, as well as other techniques. It should enable collaboration and the sharing of results with other users.

From a user's perspective, a dashboard should only contain concise, actionable information and avoid all extraneous data and visuals. If no action is required or expected from the user, the business information presented ultimately serves little or no purpose.

Presentation and Display Requirements

Two fundamental principles should guide the selection of the ideal display medium for a dashboard. It must be the best way to display a particular type of information that is commonly found on dashboards. It should be able to serve its purpose even when sized to fit into a small space.

Appropriate KPIs, charts, and colors should be used as they provide the necessary overview of the state of the business and, also, drive attention to where the performance falls short of expectations.

Bar charts are common and useful because they are relatively intuitive. Heat maps are becoming more popular since they provide a better overview of more concurrent data dimensions. GIS maps are also popular as they are intuitive and reduce the risk of misinterpreting geographical business information.

Output and Distribution Requirements

A dashboard should support the export of data, which can range from spreadsheet downloads to PDF and email outputs. The output type and quality, as well as the distribution requirements, can vary significantly. They can range from online viewing to sophisticated printouts used for external reports.

Data to be displayed should primarily be based on supporting the user's workflow rather than page aesthetics. During the requirements-gathering process, the key questions that need to be asked include the following:

- What is the likely flow of a user's focus?
- Is there a logical grouping scheme?
- Will users like to compare data? If so, then comparative data should be placed side by side.

Determining Requirements

The questions to ask for determining requirements can be grouped into the following categories:

- Objective
- Metrics
- Responsibilities and user actions
- Views/reports
- Monitoring frequency
- Security

Objective Questions

The first step requires the determination of the dashboard's objective—strategic or operational. Questions to be asked include the following:

- What are the future goals and anticipated use of the dashboard?
- Who will be the primary dashboard users? Will they require analytics or only operational information?
- Who will be the primary targets? CEO, executives, managers, department heads, line workers, or a combination of different user types?
- Will data views need to be customized for different users, across hierarchies, and across functions?

Metrics Questions

The objective of this set of questions is to determine if the processes being measured are clearly defined and if the process/system is transparent from top to bottom, with a clear breakdown of tasks, events, and responsibilities. Key questions to ask for determining the metrics include the following:

- How are effective metrics defined?
- Which metrics are essential?
- Which metrics are nice to have?
- What are the top five to ten metrics that need to be measured on a regular basis?
- How do these metrics enable informed decisions?
- Are the measures clearly defined?
- Are the dependency/relationships among different measures well understood?

Responsibilities and User Action Questions

The objective of this set of questions is to provide clarity regarding usage, responsibilities, and user actions. Key questions to ask include the following:

- How will the dashboard be used?
- Who is the business process owner?
- Who is responsible for managing or monitoring the task, events, and processes?
- What action needs to be taken upon viewing the high-level information?
- Should alerts be provided for exceptions or major issues?
- Should request for more information or status checks be automated when certain thresholds are reached?

Views/Reports Questions

The requirement for views will be different across different business units, departments, groups, and users. Any conflicting requirements will need to be addressed before the requirements are finalized. The key questions pertaining to views include the following:

- Will the dashboard communicate the complete picture?
- Will any missing business information leave the user with more questions than answers?
- What type of views will the dashboard provide? Will it be for analytical, strategic, or operational monitoring—with the ability to drill down to the transaction data? Or is there a need for a consolidated view?
- Which views/reports need to be supplemented by comparative and historical data?
- What is the navigation path for analyzing a process or report/view?
- What are the key takeaways from each report/view?
- Has a look-and-feel theme been selected which can be consistently applied across all reports, as using different colors to represent the same process across different reports can send a confusing message?

Monitoring Frequency Questions

The key questions that need to be answered pertaining to the frequency of monitoring (usage) are as follows:

- How frequently will the dashboard be monitored?
- How frequently will data be updated?
- Is the monitoring frequency in sync with the rate at which information changes and the speed at which a response must be made?

Security Questions

Security, pertaining to the application and infrastructure, is a requirement whose importance is increasing day by day. The following questions will help determine the requirements:

- How sensitive is the information being viewed?
- Who will be authorized to view the information?
- What are the role and user restrictions?
- Who will be responsible for the authorization process?
- Is there a data governance process in place?

Design Questions for User Input

User Needs

For designing a successful solution, users must be queried as they provide insights, experience, and clarity about business objectives. The key design element requires having a detailed understanding of user needs at the:

- High level (business goals, decision requirements, and workflow)
- Low level (appropriate metrics, context, and visuals)

Dashboard Design

The following is a key set of questions that will provide information pertaining to the dashboard design requirements:

- Are the appropriate KPIs being measured?
- What are the domain-specific best practice metrics for the business process?
- How actionable is the information being presented on the dashboard?
- What is the data source and what is its quality like?

- Will the dashboard provide contextual information, such as data source, refresh date/time, and schedule for the next refresh (daily, weekly, monthly)?
- How will users be empowered to act on the displayed information?
- Will users be able to add comments pertaining to the displayed information or analysis?
- Can the dashboard guide new users through a prescribed path to perform analysis, in addition to enabling them to explore data at will?
- How should data be displayed and, also, where and how does it need to be placed?
- Which types of interactions need to be supported?
- Will user collaboration be permitted?
- Will users accessing the dashboard make a decision or react to specific events or situations?
- Should some conditions, such as a critical out-of-range parameter, trigger an alert?
- How frequently will users visit the dashboard?

Placing Data

The placement and sizing of individual items on a dashboard can have a significant impact on its effectiveness. Care should be taken in deciding where and how the data will be displayed. The data to be displayed, and its location, should be based on supporting the user's workflow rather than page aesthetics. The questions that need to be asked include the following:

- What are the critical must-see or must-do items? These should be provided with prominent placement and stronger visual treatment.
- What is the likely flow of a user's focus?
- Is there a logical grouping scheme?
- Will users like to compare data? If so, then comparative data should be placed side by side.

Types of Interactions

Since a dashboard needs to support some user goals or interaction, the key questions to be asked include the following:

- What will be the users' subsequent actions, which the dashboard design will need to support?
- Which data will the users like to drill down to?
- Will the dashboard support the next step after a problem has been identified, such as opening a help desk ticket or generating an alert?
- Will users need conceptual information, such as a description or definition of an out-of-range metric?

Summary

This chapter discussed dashboard objectives and users, as well the key groups involved in their design and development. It explained, in depth, dashboard functional requirements including the key user requirements. It also discussed the presentation and display requirements, as well as the output and distribution requirements. Furthermore, the chapter explained how to determine the dashboard requirements and the questions to ask for that purpose, based on different categories, such as metrics, views, and security.

CHAPTER 5

Dashboard Design

Design Principles

Need for Effective Design

Dashboards and data visualization are cognitive tools that improve the "span of control" over business data. They help to visually identify trends, patterns, and anomalies, and guide toward effective decisions. With dashboards, scorecards, and other visualization tools now widely available to business users for analyzing their data, visual information design has become extremely important. Effective design leverages visual capabilities and is critical for clearly communicating key information to users and also for making supporting information easily accessible.

Features Affecting Dashboard Design

Six features determine how dashboards will be used and by whom. They affect the way dashboards are designed. These features are:

- Update frequency: daily, hourly, or real time
- User expertise: Novice, journeyman, expert
- Audience size: Single user, multiple users with the same requirements, multiple users who need to monitor different datasets
- Technology platform: Desktop, laptop, online, mobile device
- Screen type: Extra-large screen, standard screen, small screen, variable screen
- Data type: Quantitative, non-quantitative

A. Khan, *Visual Analytics for Dashboards*, https://doi.org/10.1007/979-8-8688-0119-8_5

Key Design Principles

The following are key design principles that must be considered when designing dashboards:

- Space: Space available on a dashboard is limited and must be used wisely.
- User relevance: Data presented must be relevant to the users' role.
- Data latency: Frequency of data updates must be relevant; while daily, weekly, or monthly data meet executive users' needs, users in operations may need higher-frequency updates, including real time.
- Personalization: Includes menus, capabilities, and custom interface.
- Interconnection with other BI tools: Dashboards are useful as a starting interface for connecting to ad hoc queries, data mining, and other more advanced BI tools.
- Collaboration tools: Ability to share a dashboard view within the enterprise or externally.
- Drill-down: Dashboard should provide the capability to act, on the signals it provides, by allowing the user to drill down to the root cause and determine a suitable course of action.
- Agility: Changing the dashboard KPIs as corporate objectives change.
- Value: Dashboard content must be more than nice to have—a dashboard must deliver actionable insight into everyday decision-making.
- Overall: Dashboard should act as a relevant indicator of performance.

Sound dashboard and visualization design will provide the business with benefits derived from improved performance. Designers should understand the users, what they need, and their business goals. They should iterate through sketches, mockups, and prototypes to explore, evaluate, and narrow down prospective solutions. They should use creativity and expertise (internal/external) to get the best ideas and the right results.

Checklist for Designing Dashboards

Stephen Few offers the following checklist for designing dashboards:

- Organize the information to support its meaning and use.
- Maintain consistency for quick and accurate interpretation.
- Make the viewing experience aesthetically pleasing.
- Design for use as a launch pad.
- Design for usability.

Impact of Poor Design

Access to appropriate and clear information is needed by every information user, including executives, analysts, managers, and knowledge workers. Poor design frequently leads to ineffective displays, which are common due to the variety of available charts and graphs, as well as a lack of training in graphics design.

As an example, graphics should clearly indicate what is being conveyed, such as target and actual. In Figure 5-1, the gauge does not provide an explicit indication of the target measure. It also uses a lot of color and space to indicate very little information.

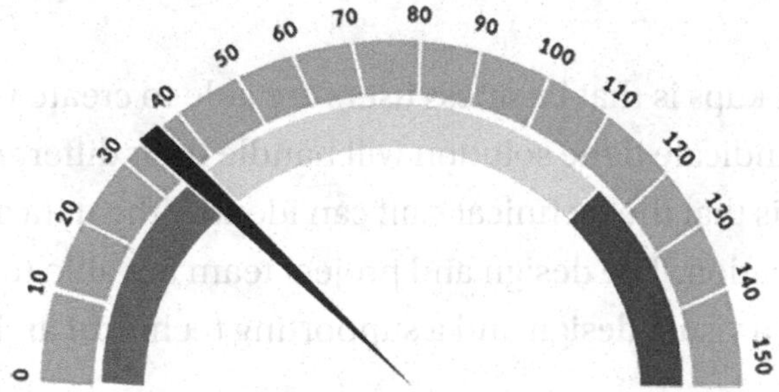

***Figure 5-1.** Inadequate gauge*

What Designers Need to Understand

A designer needs to understand the information that consumers want to see. The designer also needs to be aware of the context that each metric requires to make it useful, such as target, variance, trend, and breakdown by region. There is also a need to understand which visual representation best communicates the metric. For example, should the visual be a gauge, table, or bar chart with a reference line? Other options could be a pie chart, column chart, scatterplot, heat map, or some entirely unique visual.

Getting the right design for the visual interface is critical for success. A poor interface can have a negative impact on usage. A glitzy dashboard may obscure or miss key information.

Sketches, Mockups, and Prototypes

Sketches, mockups, and prototypes are used to conceptualize possible solutions. The design process can be significantly accelerated through a simple tabular list of metrics and indicator icons, as well as visualizations. These can reduce the risk of delivering the wrong solution.

Sketches can be as simple as whiteboard drawings or PowerPoint mockups. A mockup can be created by the business users after a whiteboard session. With a simple drawing, everyone can articulate what they need to see and also how they would like to interact with it.

The benefit of mockups is that business users are able to create various walk-through scenarios, which can indicate if the solution will handle their different requirements. An additional benefit is that the technical staff can identify the data and infrastructure required for the application. The design and project team are able to refine these whiteboard ideas into a visual design and a supporting technical architecture.

Design for Uniqueness

All dashboards are unique, which the design should reflect. Each company is defined by its culture, strengths, weaknesses, and competitive advantage, which are reflected in its unique processes, the data it captures, as well as its expertise and capabilities. Therefore, it is unlikely that an off-the-shelf dashboard will meet all the needs of a complex organization or complex system, such as an ERP or CRM system, each with its own processes and proprietary way of operating the business.

Leveraging External Expertise

If in-house dashboard expertise is lacking, it makes sense to hire consultants. Creativity, whether generated internally or with the help of consultants, potentially opens up new, innovative solutions. Individuals and organizations with expertise in dashboard design and implementation bring depth of understanding, experience of successes and failures from previous projects, as well as research and diversity of skillsets.

Architecture

How to Architect Dashboards

There exist many ways in which dashboards, which are full-fledged information systems, can be architected. Such a system requires extracting and merging data from multiple systems. Its technical architecture must map to the business if the desired functionality is to be delivered.

The first step in the design process is to ask three basic questions, which will identify the specific reason why the dashboard will be useful to the organization:

- Who is my audience?
- What value will the dashboard add?
- What type of dashboard needs to be developed?

Business and Technical Architecture

The designer should understand the organization's business architecture, which consists of stakeholders (investors, board, executives, workers), strategy (mission, goals, objectives, vision, values), as well as tactics and resources (people, technology, capital, projects). The performance management system should align with the organization's business architecture.

The components of the technical architecture include displays, applications, data stores, integration, and data sources.

When the architecture of a performance management system is designed, designers should select different components at each level of the architecture that best meet the business needs. For example, strategic dashboards use a scorecard interface that is periodically updated both automatically and manually from Excel files, web pages, and various applications. On the other hand, operational dashboards frequently use custom APIs, ETL, EII (Enterprise Information Integration), and caching technologies to pull data from legacy and other transaction systems.

The key to joining the business and technical architectures is metrics, which accurately reflect and measure the business strategy and performance.

Effective Principles of Dashboard Design

Avoid Information Overload

The inclination of most dashboard developers is to display all kinds of information. However, it should be realized that some data can be a distraction and not relevant to the decision-making process. Information overload should be avoided as it can hinder the dashboard's effectiveness.

Design for Rapid Monitoring

Dashboards should be designed for rapid performance monitoring. They effectively support rapid performance monitoring only if designed to work with the human eyes and brain. The dashboard content must support efficient and meaningful monitoring. Important items should be made more visually prominent than less important items. Items that need to be scanned in a particular order should be arranged in a manner that supports that sequence of visual attention.

Implement Responsive Design

Appropriate design encompasses the idea of responsive design. This is an approach to web design which supports the layout of web pages regardless of the visitor's screen size and orientation. This is an important factor as no organization is immune to the BYOD phenomenon and, hence, needs to account for the variety of devices accessing the application. Responsive design in BI applications is increasingly important for supporting the various mobile devices being introduced into the corporate environment.

Provide Insight, Not Reports

Dashboard users want to use the tool to understand and run their business. What they want is insight, not static reports, which can be obtained from other sources and platforms. Therefore, the dashboard content should help them achieve their objective. This can only be done if a serious effort is made to understand what the users really want and then document the requirements.

A dashboard can be made more effective by making it interactive and providing flexibility through filters, dropdowns, column selectors, view selectors, buttons, check boxes, etc. If a user cannot derive insight in the first few seconds of viewing a dashboard, the design could use improvement.

A successful project will include designers who know the business side well. They can develop content without explicit instructions from the business users. Such designers will move away from asking what the report should look like or which data is required. Instead, they will move toward understanding the business processes and determining which data and reports will provide insight that the users need.

Utilize Valuable Real Estate

Space on a dashboard is very limited, where numerous items compete for space. Therefore, organizing the screen becomes very important. Screen size is a big limitation when it comes to dashboard design, as it is only possible to cram so much onto a typical laptop or mobile device screen. White (blank) space should be eliminated as it projects an incomplete look and also wastes valuable real estate.

Avoid Scrolling

On a dashboard, scrolling should be avoided. Therefore, it creates a challenge to design a screen where everything that users need can be fitted. This can be approached by thinking of dashboards as websites with navigation and multiple screens, where the first screen is a summary dashboard, a condensed version of the user's most important KPIs. The subsequent dashboard views can present more details and other views.

Use an Iterative Design Process

Dashboard design requires multiple iterations to be successful. Business users often don't know or realize what they want until they can see something they like or recognize something that needs to be changed. Therefore an iterative process should be used. The development plan should include several refinement iterations. The plan should include a strategy for capturing and documenting user input, prioritizing changes, and determining complexity. Careful planning of this process will ensure that all iterations run smoothly and add as much value as possible. An iterative process will ensure that there are no surprises and, therefore, the final product will quickly get accepted by the users.

Implement Best Practices and Standards

To develop the best and most effective dashboards, best practices should be followed. These can be quickly implemented on every new dashboard. Dashboards implemented while following best practices have received overwhelming user acceptance. The approach of getting something out the door and then going back later to apply best practices should be avoided. Standards and best practices should be used from the beginning, starting with the first release.

Layout, Colors, and Fonts

Layout

Dashboard layout is an important aspect of dashboard design. How a dashboard works and how the user should interact with the dashboard should take precedence over how the dashboard is going to look and feel. A good designer will be successful in finding a balance between the two.

Chart Elements

The most common type of visualization on a dashboard is the chart, whose key elements are shown in Figure 5-2.

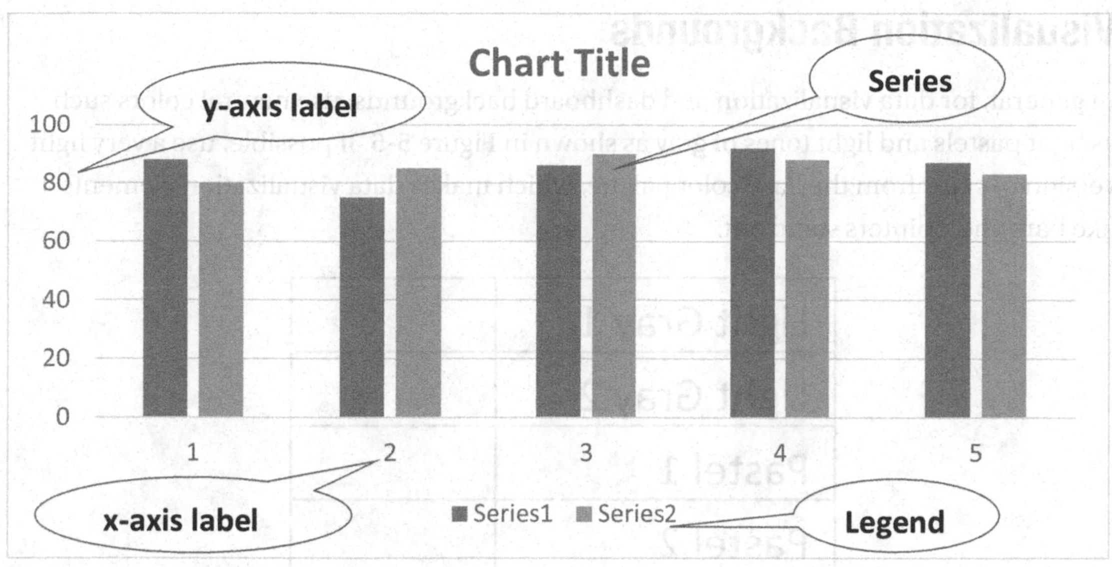

Figure 5-2. *Chart elements*

Colors

Before starting any styling, the starting color palette should be identified. Usually, there are corporate colors to work with. If there are no brand colors or logos, generally go with a simple palette of white, a primary color like a cool blue, and a light gray as shown in Figure 5-3.

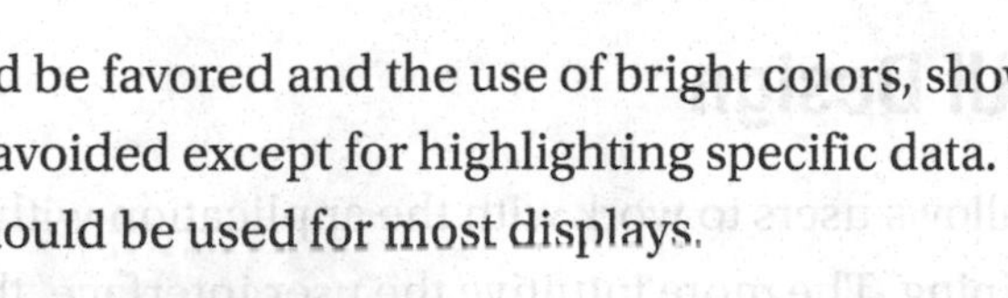

Cool Blue	
Light Gray	

Figure 5-3. *Neutral colors*

Light colors should be favored and the use of bright colors, shown on the left in Figure 5-4, should be avoided except for highlighting specific data. Subdued colors, shown on the right, should be used for most displays.

Figure 5-4. *Bright versus subdued colors*

Visualization Backgrounds

In general, for data visualization and dashboard backgrounds, use neutral colors such as light pastels and light tones of gray as shown in Figure 5-5. If possible, use a very light version of color from the base color palette, which makes data visualization elements like bars and pointers stand out.

Light Gray 1	
Light Gray 2	
Pastel 1	
Pastel 2	

Figure 5-5. *Neutral background colors*

Fonts

Attention to detail in dashboard design is important, especially for fonts. Use consistent font types and sizes throughout the entire dashboard. In a web environment, Trebuchet MS or Verdana is recommended for desktop dashboards. For titles, use Myriad, Calibri, or Arial, while for the content use Verdana or Tahoma.

Decide early on the minimum font size for the dashboard text. Font sizes for objects like axis titles, axis scales, and legend descriptions are around 10pt. Use 12pt to 14pt and bolded text to distinguish the main titles from the other titles.

UI Design Principles

Importance of UI Design

A good user interface allows users to work with the application without having to read manuals or receive training. The more intuitive the user interface, the easier and less expensive it is to use. The better the UI, the easier it is to train people to use it, which reduces the training cost. Also, less help will be required by the users, which will reduce the support costs.

User/UI Interaction

The fundamental reality of application development is that the user interface is the system for the users, who want applications that meet their needs and are easy to use. However, in many cases, that is not what they get.

A common UI design issue is the developer issue. Too many developers think that they are artistic geniuses and do not bother to follow UI design standards or invest the effort to make their applications usable. They mistakenly provide importance to items that are not important from a user's perspective, such as implementing an eye-catching color scheme.

Principles of UI Design

There are six basic principles of user interface design:

- Structure principle
- Simplicity principle
- Visibility principle
- Feedback principle
- Reuse principle
- Tolerance principle

Structure Principle

The structure principle is concerned with the overall user interface architecture. The design should organize the UI in meaningful and useful ways based on clear and consistent models, which are apparent and recognizable to users. Related items should be grouped together, while unrelated items should be separated. The design should help differentiate dissimilar items, while making similar items resemble one another.

Simplicity Principle

According to this principle, the design should make easy and common tasks simple to perform. The communication should be clear and simple in the user's own language. Good shortcuts should be provided, which are meaningfully related to longer procedures.

Visibility Principle

According to the visibility principle, the design should keep all needed options for a given task visible without distracting the user with extraneous or redundant information. Good designs do not overwhelm users with too many alternatives or confuse users with unnecessary information.

Feedback Principle

The feedback principle requires that the design should keep users informed of any actions or interpretations, changes of state or condition, as well as errors or exceptions that are relevant and of interest to them. The communication should be through clear, concise, and unambiguous language that users are familiar with.

Reuse Principle

According to the reuse principle, the design should reuse internal and external components and behaviors. Also, the design should maintain consistency.

Tolerance Principle

According to the tolerance principle, design should be flexible and tolerant, reducing the cost of mistakes and misuse by allowing undoing and redoing. Wherever possible, errors should be prevented by tolerating varied inputs and sequences and, also, interpreting all reasonable actions.

Summary

This chapter delved into dashboard design, stressing the importance of effective principles in enhancing data visualization and decision-making. It outlined six key features, including update frequency and user expertise, which affect the way dashboards are designed. It described key design principles and provided a checklist for designing dashboards. The chapter explored the impact of poor design and also discussed the architecture, business and technical. Finally, it described the effective principles of dashboard design, as well as layout, color, font, and UI design principles.

CHAPTER 6

Data Visualization

Memory and Cognitive Load

Memory Types

When light is reflected off of a stimulus, it is captured by the eye. However, most processing is done by the brain, which we think of as visual perception. The brain has three types of memory:

- Iconic memory
- Short-term memory
- Long-term memory

Each type of memory plays an important and distinct role.

Iconic Memory

This type of memory is super-fast and is not realized by the person when it is triggered. Iconic memory comes into play when we look at the world around us. As humans developed, they developed the ability to quickly recognize differences in the environment, such as an animal's movement, which became ingrained in their visual process. These primitive survival mechanisms can now be leveraged for effective visual communication.

Information stays in the iconic memory for a fraction of a second before it gets forwarded to the short-term memory. Iconic memory is tuned to a set of pre-attentive attributes, which are critical tools for visual design.

A. Khan, *Visual Analytics for Dashboards*, https://doi.org/10.1007/979-8-8688-0119-8_6

Short-Term Memory

Short-term memory, also known as primary or active memory, is the capacity to store a small amount of information in the mind and keep it readily available for a short period. Humans can keep about four chunks of visual information in their short-term memory at a given time. The duration of short-term memory is estimated to be seconds.

Long-Term Memory

Long-term memory is the storage system that enables individuals to retain, retrieve, and make use of knowledge hours, weeks, or even years after such information has been learned. It is the final stage in the processing of memory. When something leaves the short-term memory, it is either lost or goes into long-term memory, which is built over a lifetime. It is very important for pattern recognition and general cognitive processing.

Cognitive Load

In the business world, it is not unusual to view a presentation slide and wonder what it was all about, due to it being too busy and/or complicated. Cognitive load occurs whenever information is taken in. It is the mental effort required to learn new information. Since the human brain has a finite amount of mental processing power, the information designer needs to be smart about how to use the audience's brain power. Therefore, the visualizations created should avoid extraneous cognitive load processing, which takes up mental resources but does not help in understanding the information. The perceived cognitive load should be minimized to a reasonable extent, while still allowing information to get across to the audience.

Need to Visualize Data

Memory Limitations

To bypass short-term memory limitations, humans tend to reduce the need to rely on it. In the business environment, one way to do that is to place all the important information on a single screen.

A dashboard acts as an external form of memory where all information is available in front of the eye. It takes away the need to memorize information, that is, to move it into long-term memory, where it is retained for years. The explanation for this phenomenon is that when information is in front of the eyes, it can be quickly moved in/out of memory at an extremely fast speed as it is processed. A dashboard with a well-designed layout and content facilitates rapid processing.

Key Characteristics of Good Visualizations

Here are the key characteristics of good visualizations:

- Fitted to audience
- Fitted to purpose
- Accurate
- Distortion free
- Precise
- Clear
- Understandable
- Visually appealing
- Communicates knowledge
- Encourages comparisons
- Tells a story
- Inspires conversations

Data Visualization in Business

Effectiveness of Visuals

Humans primarily perceive their surroundings through their eyes, despite the availability of five sensory skills which allow them to see, touch, hear, taste, and smell. Seeing equates to understanding because vision dominates the senses and is faster.

It is well understood and appreciated that a picture is worth a thousand words, which explains why visualizations like charts are more powerful than tables. When something is displayed on a chart, the data gets highlighted immediately. Therefore, since data is presented on a dashboard in a visual form, it becomes a very effective medium for conveying information.

Comparing Charts with Tables

Why Charts Are More Powerful Than Tables

Table 6-1 displays call volumes for two call centers, US and Europe, over a period of 12 months (24 values).

Table 6-1. *Tabular display of call volume data*

	Call volume (in thousands)											
	Jan	Feb	Mar	Apr	May	Jun	Jul	Aug	Sep	Oct	Nov	Dec
US	2,975	3,515	3,890	3,425	3,861	4,257	3,573	3,951	4,407	4,109	4,475	5,225
Europe	861	954	1,010	1,501	966	1,019	890	209	899	875	903	1,035
	3,836	4,469	4,899	4,926	4,827	5,276	4,463	4,160	5,306	4,983	5,378	6,260

After scanning the table of numbers, once the eyes are taken off of it, at the most only four or five numbers will be remembered—one per slot of memory. The same data is represented as a chart in Figure 6-1.

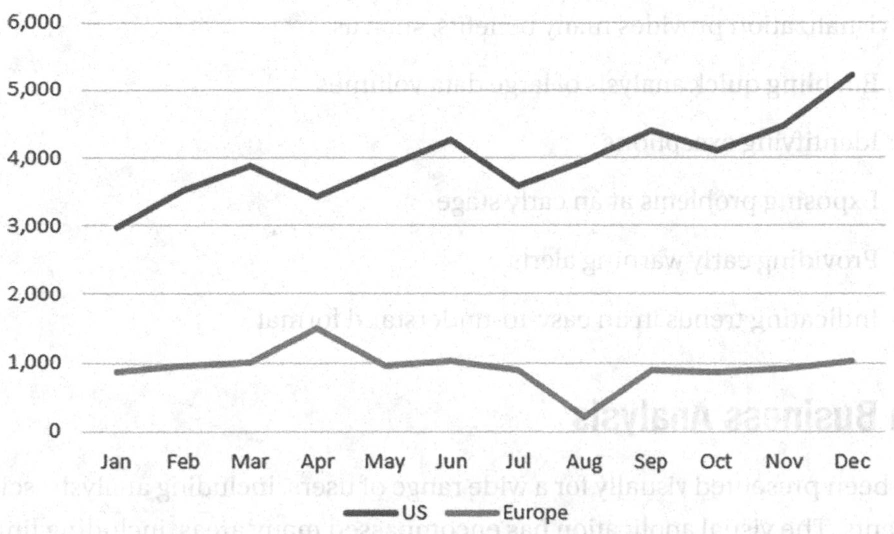

***Figure 6-1.** Chart data highlights patterns and trends for Table 6-1 data*

When the line chart is viewed, the entire pattern formed by the line's shape, that is, 12 monthly values, can be stored as a single chunk of memory. When quantitative information is displayed in visual form, much more information can be stored in the short-term memory compared to values written as numbers. The chart, simple to analyze, displays the values meaningfully as a line, which goes up and down to represent the rise and fall of values over time.

Data Visualization for Business

Benefits

The foundation of successful businesses is sound and informed decision-making, which is based on data. The problem is that as an organization grows and the available data grows in tandem, it becomes a challenge to analyze such data. This is especially true if provided in spreadsheets or static reports, which do not facilitate the determination of data patterns and trends. Decision-makers in such organizations need BI tools, especially visual tools, which can help them make decisions quickly and correctly.

Data visualization provides many benefits, such as

- Enabling quick analysis of large data volumes
- Identifying exceptions
- Exposing problems at an early stage
- Providing early warning alerts
- Indicating trends in an easy-to-understand format

Role in Business Analysis

Data has been presented visually for a wide range of users, including analysts, scientists, and students. The visual application has encompassed many areas, including finance and astrology. However, as a presentation tool, its widespread use and importance in business is a recent phenomenon. Data visualization has been driven by the large volumes of data being collected, which is increasing exponentially, and the widespread availability of BI technology and analysis tools at an affordable cost.

The presentation medium of choice for fast and accurate decision-making has been the dashboard. It is now considered by businesses to be an important tool for remaining competitive. Despite this fact, data visualization value is often not sufficiently appreciated or is implemented ineffectively.

Key to Successful Data Visualization Usage

Developing various visualization types is very easy and can be done quickly using modern data visualization tools like Tableau and Power BI. However, for maximum effectiveness, the key is to use the most appropriate type of data visualization. Also, the data presented should be supplemented with contextual information that provides clarity and enables users to take action.

Presentation Media

Presentation Media Types

Four types of presentation media are widely used in business:

- Dashboards
- Scorecards
- Visual analysis tools
- Reports

All four types have their own unique attributes, which help users to evaluate the state of their business and identify trends, patterns, correlations, anomalies, and deviations.

On a dashboard, data can be presented using a wide variety of visualizations, with different presentation characteristics.

Visual Display Elements Used by Dashboards

Dashboards use a variety of visual display elements. The commonly used display elements include charts, graphs, maps, gauges, annotations, tables, and visual icons.

Summary

This chapter explored data visualization, focusing on memory, cognitive load, and the role of effective visualizations in business. It discussed three memory types—iconic, short-term, and long-term—and their significance in visual perception. The chapter emphasized minimizing cognitive load in visualizations to enhance comprehension and highlighted the benefits of dashboards as external memory aids. Key characteristics of good visualizations were outlined, alongside their superiority over tables in data retention. Additionally, the chapter explored the benefits of data visualization in business decision-making and the importance of choosing appropriate visualization types.

CHAPTER 7

Visualization Principles

Attributes of Visual Perception

Pre-attentive Visual Attributes

Pre-attentive processing is a term that describes how people respond to visuals as a result of visual hierarchy. Pre-attentive processing, visual perception that occurs below the level of consciousness, should be used when designing charts.

Colin Ware, author of *Information Visualization: Perception for Design*, has described 17 pre-attentive attributes of visual perception, which can be organized into four categories. These are color, form, position, and motion. Examples of the first three categories, which are most commonly leveraged, are shown in Figure 7-1, where they are represented as follows:

- Color: Hue and intensity
- Form: Line length, line width, orientation, size, shape, added marks, and enclosure
- Position: Spatial position

Motion can be represented by flicker, where the object repeatedly appears and disappears.

A. Khan, *Visual Analytics for Dashboards*, https://doi.org/10.1007/979-8-8688-0119-8_7

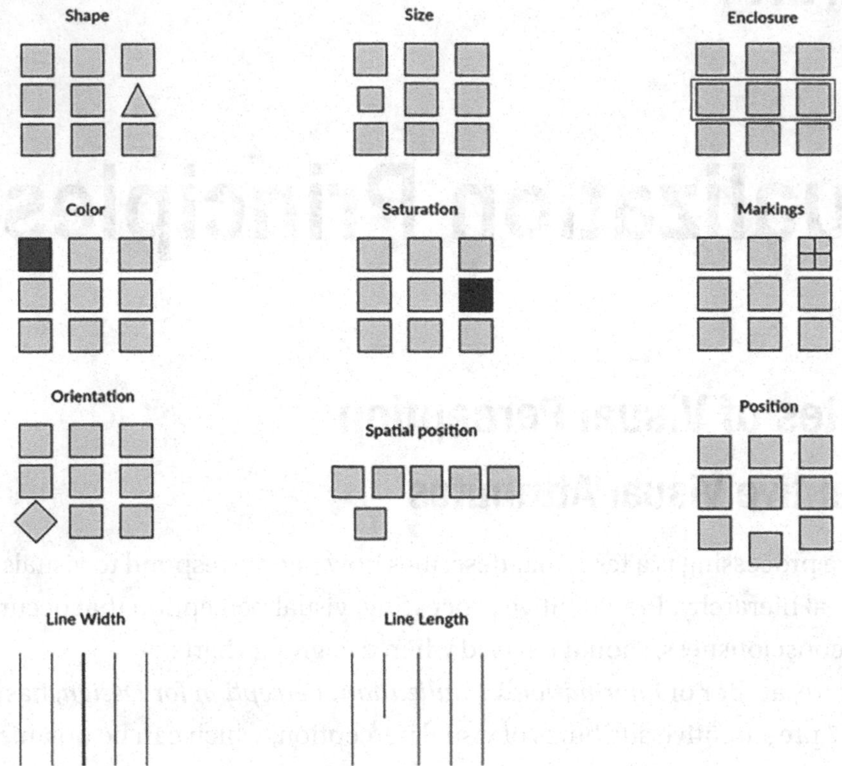

Figure 7-1. *Pre-attentive attributes*

Leveraging Pre-attentive Visual Attributes

Pre-attentive processing of visual information is performed automatically on the entire visual field, while detecting the basic features of displayed objects. The basic features include colors, closure, line ends, contrast, tilt, curvature, and size. These simple features are extracted from the visual display in the pre-attentive system and, later, joined in the focused attention system into coherent objects.

Pre-attentive visual attributes can be strategically used in two ways:

- Leverage to help direct the audience's attention to where the designer wants them to focus.
- Create a virtual hierarchy of elements to lead the users through the information which the designer wants to communicate and, also, in the way they should process it.

Visual designers exploit visual hierarchy and pre-attentive processing using techniques such as:

- Varying the shape
- Enclosing a group
- Changing line thickness
- Changing object size
- Applying markings to an object
- Changing an object's orientation relative to the visual space
- Adjusting saturation or hue
- Changing an object's position relative to similar objects

By understanding how the audience views and processes information, a designer will be in a better position to communicate effectively. In Figure 7-1, when scanning the attributes, the eyes are drawn to the one element within each group that is different from the rest. One does not have to look for it because the brain is hardwired to quickly pick up differences it notices in the environment.

When used sparingly, pre-attentive attributes can be extremely useful for drawing the audience's attention quickly to where they need to look and for creating a visual hierarchy of information.

Four Categories of Pre-attentive Attributes

Color Attribute

Color, which is a very powerful tool when used properly, is defined by three attributes:

- Hue
- Saturation
- Lightness/brightness

Hue is represented by various colors like red, blue, and green. Saturation indicates the intensity or purity of a color. As saturation increases, the colors appear to be purer. Saturation is represented as a percentage, where 100% is full saturation and 0% is a shade of gray, as shown in Figure 7-2. In other words, saturation indicates the percentage of hue where a value of 0% means hue is absent, while a value of 100% means that hue is fully present.

Color saturation determines how a particular hue will look in certain lighting conditions. For example, an object painted in a solid color will look different during the day compared to what it will look like at night. The object saturation will change as the day progresses due to the light, even though it is the same color.

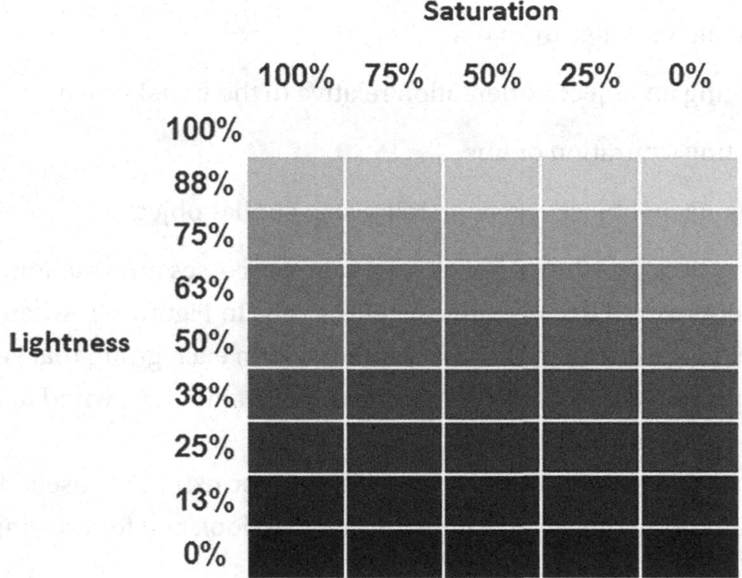

***Figure 7-2.** Saturation versus lightness*

Brightness, also called lightness, refers to how dark or light the color is and is represented as a percentage. Brightness is the relative lightness or darkness of a particular color, which can range from black (no brightness) to white (full brightness). It measures the degree to which any hue appears dark or light, ranging from fully dark (black) to fully light (white). A value of 0% means the color is fully dark and appears to be black, 50% is normal, while a value of 100% means the color is fully light and appears to be white. Lightness and saturation together are also referred to as intensity.

Color can be significantly influenced by the surroundings or background, as shown in Figure 7-3, where the text in the third box is not visible due to the similar background color.

***Figure 7-3.** Effect of background color*

Form Attribute

The visual attributes of form can be used to make dashboards more effective. For example, line length can be used in bar graphs for encoding values. Similarly, line width can be used for highlighting by increasing the line thickness. Other forms commonly used are orientation and size. The relative size of an object can be used to increase or decrease the importance of various items such as charts, tables, titles, etc.

Added marks are often used on dashboards, typically as simple icons, by placing them next to the item that needs attention. These marks can be a simple "x," an asterisk, a circle, etc. The form attribute enclosure is used for grouping items or sections.

Position Attribute

The position category of pre-attentive visual attributes includes both the two-dimensional position of the data in the visualization and the spatial grouping of the data points. The two-dimensional position can be visualized where, on the image, a data point or points reside (low, high, off to the side, etc.). For charts, the most widely used means for encoding quantitative data is the 2-D pre-attentive attribute.

Motion Attribute

The fourth category of pre-attentive visual attributes is motion, which consists of flicker and direction of motion. Flicker is a very effective tool for getting attention, as something appearing and disappearing alternatively is a powerful attention grabber. While flicker is a powerful tool, its use should be minimized, as it can lead to the user getting annoyed or ignoring it. The motion attribute can be useful when attention needs to be drawn to a certain trend or insight.

Gestalt Principles of Visual Perception

Gestalt Principles

There are six principles of visual perception, which are also known as the Gestalt principles. They define how people interact with and create order out of visual stimuli. The six principles are:

- Proximity
- Similarity
- Enclosure
- Closure
- Continuity
- Connection

There are two types of elements which can exist in a visual:

- Signal: Information that needs to be communicated
- Noise: Clutter that needs to be avoided

The Gestalt principles of visual perception help to identify elements that are signal and noise.

Proximity Principle

When objects are physically located near each other, they are perceived to be part of a group. Figure 7-4 shows that the dots are three distinct groups because of their proximity to each other.

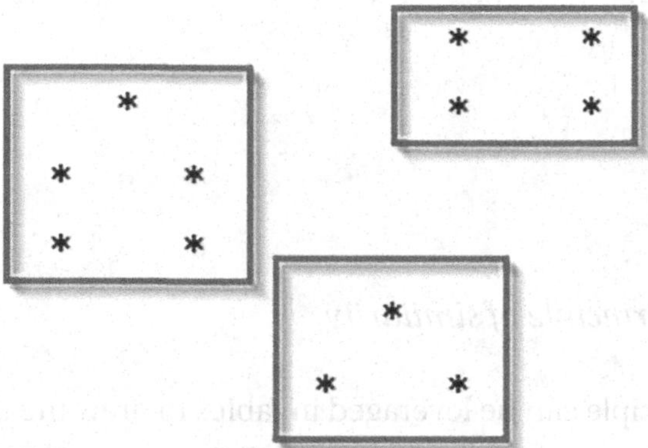

Figure 7-4. *Gestalt principle of proximity*

The proximity principle can be used to direct viewers in a particular direction, such as horizontal or vertical. In Figure 7-5, by differentiating the space between the dots, the eyes are drawn either

- Down the columns (first set)
- Across the rows (second set)

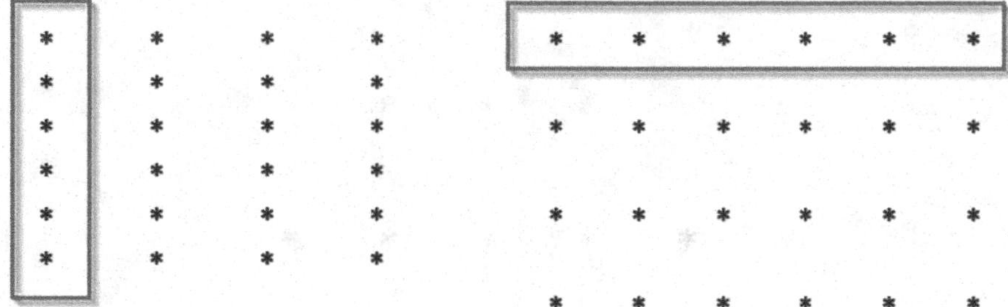

Figure 7-5. *Effect of spacing dots*

Similarity Principle

Objects that are similar in size, color, shape, or orientation are perceived to belong to a group. In Figure 7-6, the blue asterisks are associated together (on the left), just like the red triangles (on the right).

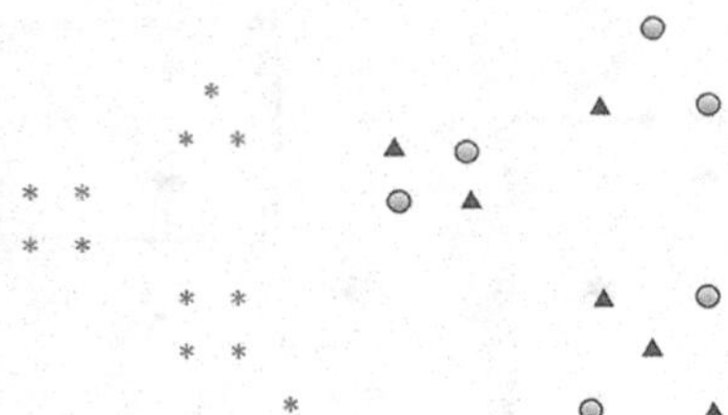

Figure 7-6. *Gestalt principle of similarity*

The similarity principle can be leveraged in tables to draw the audience's eyes in the direction that the designer wants. In Figure 7-7, similarity of color is a cue for the eyes to read across the rows, rather than down the columns. This helps eliminate the need for additional elements, such as borders, to direct the user's attention.

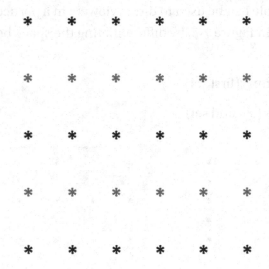

Figure 7-7. *Color similarity cue*

Enclosure Principle

When objects are physically enclosed by anything that forms a border around them, they are perceived to belong together, as shown in Figure 7-8. The enclosure makes the object within stand apart from objects that are outside it. The enclosure does not need to be a strong one, such as a thick line. Frequently, light background shading is often sufficient to create the enclosure.

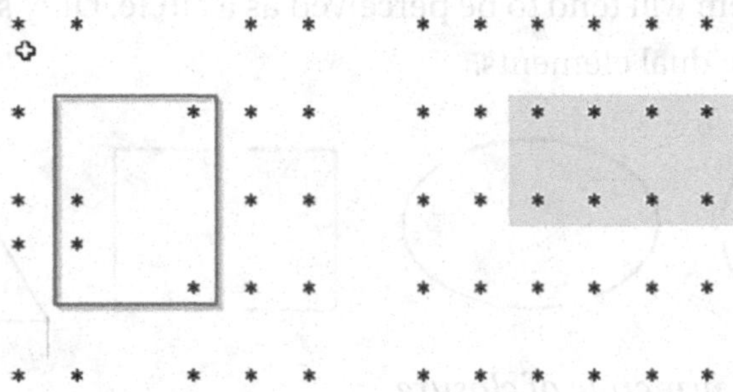

Figure 7-8. Gestalt principle of enclosure

Figure 7-9 shows how the enclosure principle can be leveraged to draw a visual distinction within the data.

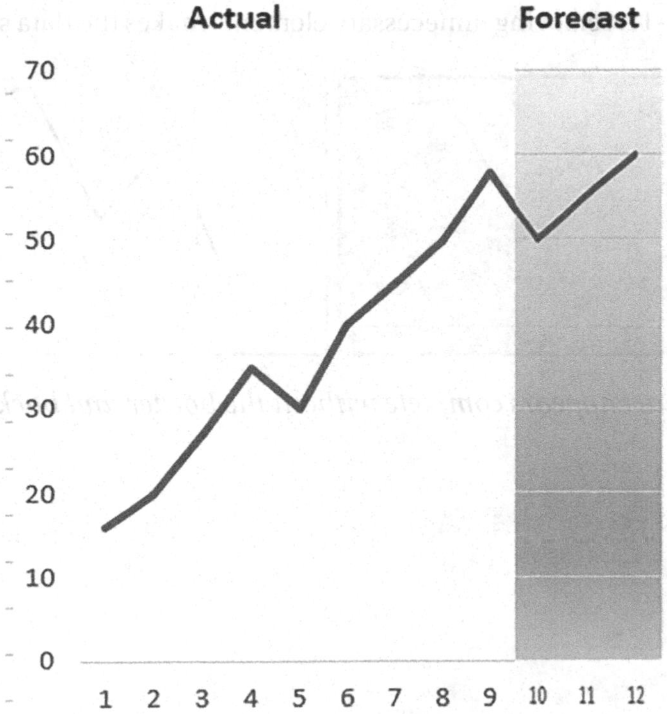

Figure 7-9. Chart leveraging the enclosure principle

Closure Principle

Humans like things to be simple and to fit in the constructs that exist in their heads. According to this principle, open structures are perceived as closed, complete, and regular. Therefore, there is a tendency to perceive a set of individual elements as a single recognizable shape when possible.

When parts of a whole are missing, the eyes fill in the gap. In Figure 7-10, the elements of the first item will tend to be perceived as a circle. Only subsequently will they be viewed as individual elements.

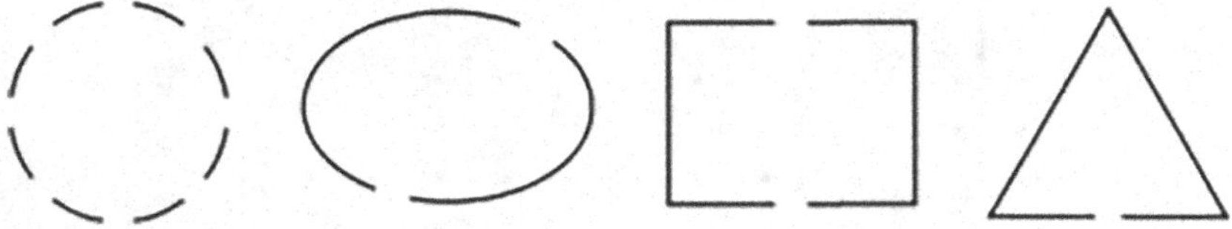

Figure 7-10. *Gestalt principle of closure*

Many applications, like Excel, have default chart settings that include elements like borders and background shadings. According to the closure principle, such elements are unnecessary. If they are removed, the chart will still appear as a cohesive entity. As shown in Figure 7-11, removing unnecessary elements makes the data stand out.

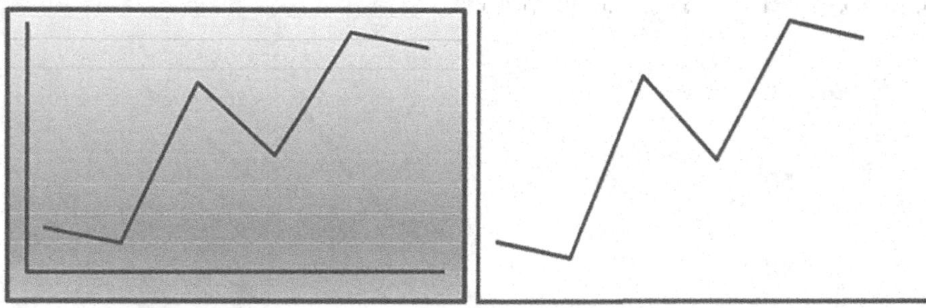

Figure 7-11. *Chart appears complete without the border and background shading*

Continuity Principle

The continuity principle is similar to the closure principle. If objects are aligned with each other or appear to continue one another, they are perceived as belonging together. When looking at objects, the eyes seek the smoothest path and naturally create continuity in what is seen, even where it may not exist explicitly.

In Figure 7-12, if the two objects of the first item are pulled apart, most people will expect to see what is displayed on the second item, even though it can be as shown in the third item.

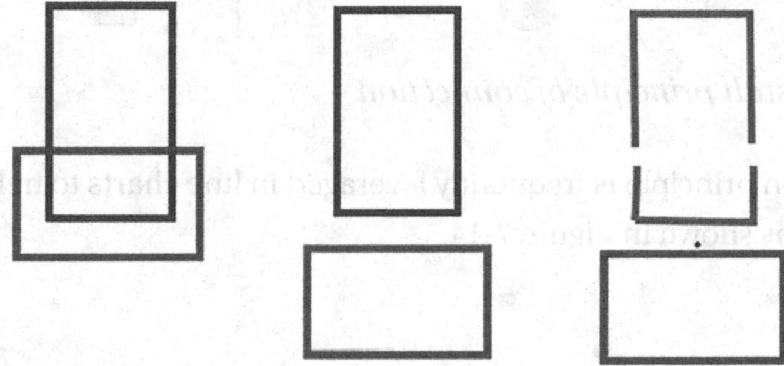

Figure 7-12. *Gestalt principle of continuity*

Connection Principle

There is a tendency to think of objects that are connected, such as by a line, as part of the same group. Connective property typically has a stronger associative value than proximity or similarity (such as size, shape, and color). However, the perception of grouping due to connection is weaker only than that produced by enclosure.

Visual perception can be manipulated through the thickness and darkness of lines to create the desired visual hierarchy. In Figure 7-13, due to the connection principle, the eyes pair the shapes connected by lines (rather than similar color, size, or shape).

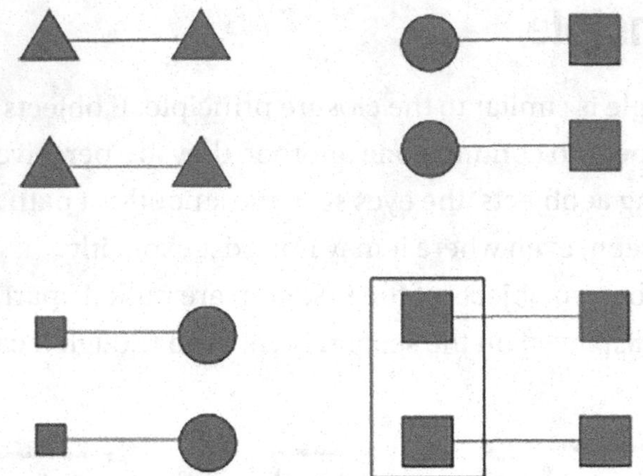

Figure 7-13. *Gestalt principle of connection*

The connection principle is frequently leveraged in line charts to help the eyes see order in the data, as shown in Figure 7-14.

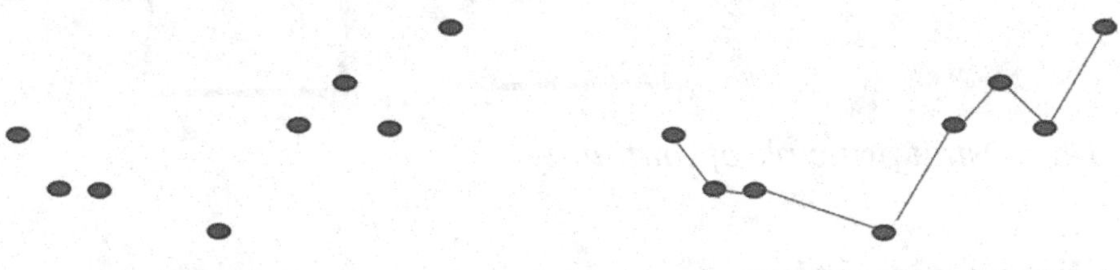

Figure 7-14. *Lines connect the dots*

Visual Design Considerations

Size

Size matters, as relative size denotes relative importance. Therefore, it should be kept in mind when designing visual communications. When showing multiple items that are of roughly equal importance, they should be sized similarly. If there is one really important item that a user needs to focus on, then size should be leveraged by making it bigger.

Color

When colors are used to represent either categorical groupings of data or quantitative data values, the following should be considered:

- Distinctness:
    - Are the different hues chosen sufficiently distinct from each other to be interpreted by the reader?
- Cultural context:
    - How will different groups interpret a variety of colors in different contexts?
    - Red may mean alarm in the United States compared to good luck in another country.
- Psychological response:
    - Use pastel or softer hues for information that does not need to be emphasized.
    - Reserve the use of bright or harsh hues to highlight something important in the chart.
- Color blindness:
    - Be aware that color blindness occurs in a percentage of users.
    - Many users cannot distinguish between green and red items.

Fonts

When selecting fonts, strive for clarity and legibility. Attempts should be made to maintain a consistent font style throughout a table, chart, or dashboard. Consistently maintain font size, bolding, and italics. Exceptions can be made for highlighting specific values. However, they should be used sparingly to maintain effectiveness.

Chart Scale and Size

Chart design should ensure that distortion, exaggeration, and manipulation are avoided. The focus should be on communicating facts. Falling into a design trap and unintentionally distorting or exaggerating a chart's true meaning should be avoided. The two areas that are susceptible to these issues are chart scale and size.

Chart Scale

The chart scale defines how the distance along one of the chart's axes should be interpreted. Quantitative scales assign numeric values, while categorical scales assign categorical divisions. The tick marks are used to show the increments along a given axis. When an arithmetic scale is used, the interval between tick marks is always the same. The lower end of the quantitative scale must be assigned a value. Choosing a value larger than zero makes it possible to exaggerate the information presented by the chart.

Figure 7-15 shows two charts with the y-axis for the first chart starting from zero, while the second chart's y-axis starts from 60. On the first chart, the first two bars, blue and red, show a value of 70 and 88 when the scale starts from zero. On the second chart, the values are the same but the scale starts from 60. However, on the second chart, it appears that the blue bar is about a third of the value of the red bar, which the first chart does not indicate. Therefore, a starting value of zero should always be used for the y-axis for bar charts.

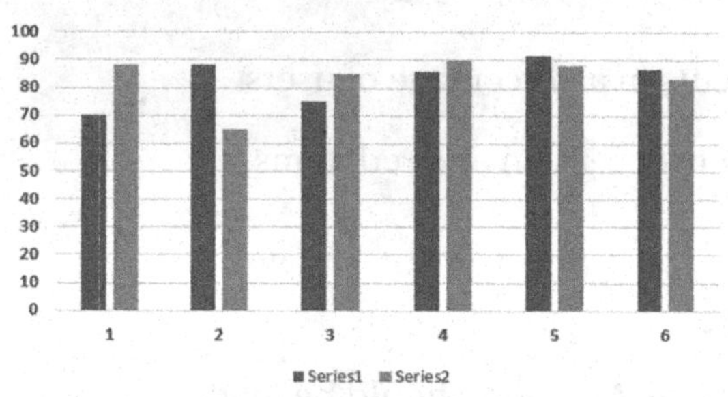

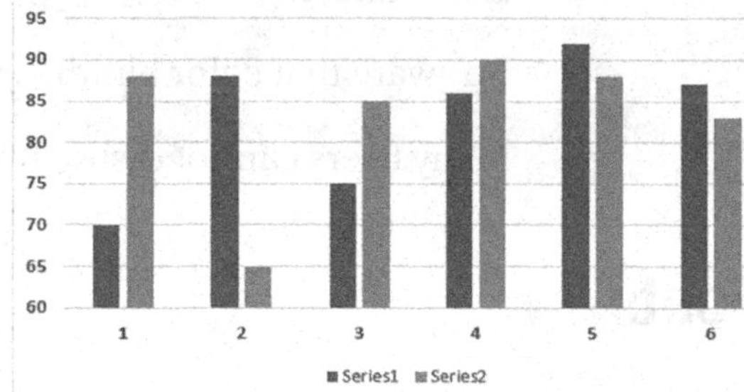

***Figure 7-15.** Different scale values*

Chart Size

The chart size is defined by its width and height. The aspect ratio is the ratio of the length of the vertical axis to the horizontal axis. The choice of aspect ratio significantly impacts how the information presented by the chart is perceived. Figure 7-16 shows how the aspect ratio can distort a visual.

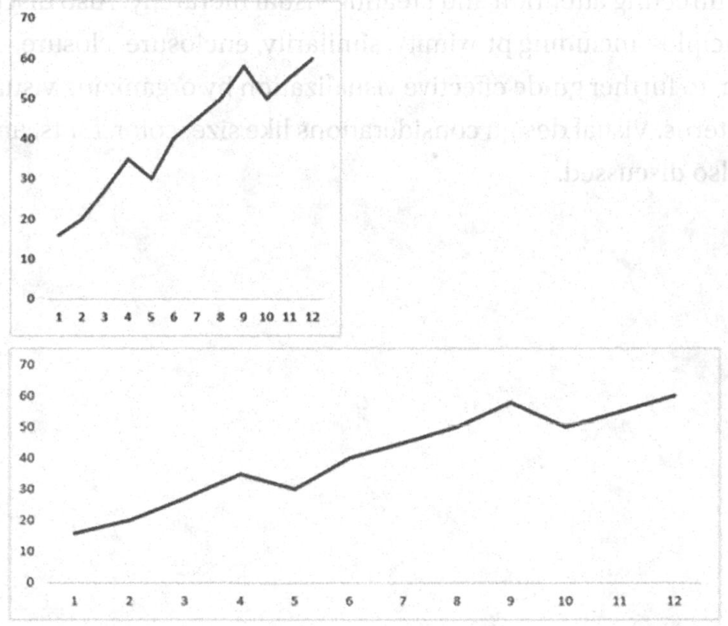

Figure 7-16. *Effect of aspect ratio*

It is generally recommended that the aspect ratio should be less than 1, which means that the area should be wider than it is tall. Aspect ratios greater than 1 tend to exaggerate how the chart is perceived.

Summary

This chapter delved into visualization principles, particularly focusing on pre-attentive visual attributes. Pre-attentive processing, occurring below conscious awareness, influences how people respond to visuals. Colin Ware's 17 pre-attentive attributes, categorized into color, form, position, and motion, were explored, highlighting their strategic use in directing attention and creating visual hierarchy. Also discussed were the Gestalt principles, including proximity, similarity, enclosure, closure, continuity, and connection, to further guide effective visualization by organizing visual stimuli into meaningful patterns. Visual design considerations like size, color, fonts, and chart scale and size were also discussed.

CHAPTER 8

Visual Communication

There are four components of visual communication:

- Visual cues
- Coordinate systems
- Measurement scales
- Visual context

Visual Cues

Visual cues draw the viewer's eyes to relationships or specific parts of a chart. These include placement, lines, shapes, and color. Visual cues need to be effective to ensure perception accuracy.

The order of the most accurately perceived visual cues is as follows:

- Position is the most accurate.
- Line properties (length, angle, and direction).
- Shape properties (area and volume).
- Color properties, saturation and hue, are least accurately perceived.
- Saturation is more accurately perceived than hue; users with a red-green deficiency can distinguish saturation even when they cannot identify the hue.

A. Khan, *Visual Analytics for Dashboards*, https://doi.org/10.1007/979-8-8688-0119-8_8

Placement

Placement locates items on a chart with reference to:

- Position
- Proximity

Position

The positioning of an item affects how viewers perceive what they see. In most languages, reading is from left to right and top to bottom. In visuals, such learning is transferred, as shown in Figure 8-1:

- In the left figure, where the eyes go first, revenue is considered the most important.
- In the right figure, where the eyes go first, profit is considered the most important.

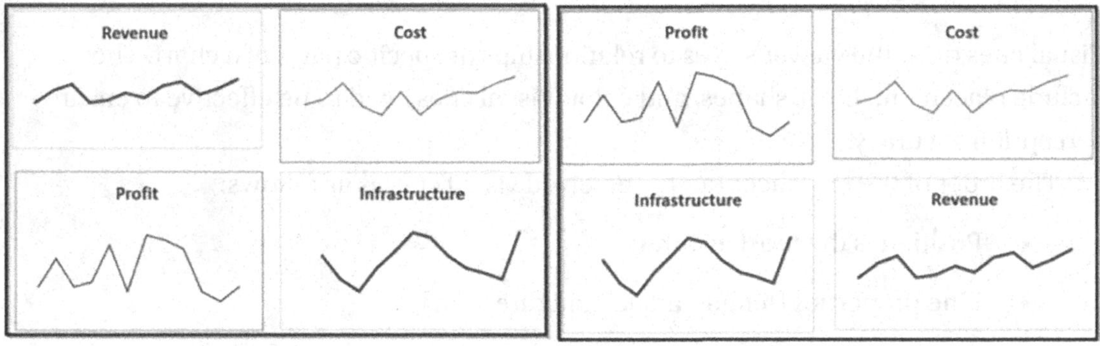

***Figure 8-1.** Position*

Proximity

Proximity affects visual perception. Items placed close together are viewed as being related, while items at a distance are viewed as being unrelated or less closely related, as shown in Figure 7-5 (Chapter 7).

Lines

The length, angle, and direction of lines on a chart directly influence visual perception. Lines are very sensitive to visual perception and distortion. The three characteristics which influence how we perceive lines are length, angle, and direction.

Length is the distance between the two endpoints of a line. Angle is the orientation of a line relative to another object (or to a real or perceived horizontal plane). It is expressed in degrees and ranges from 0 to 360 degrees. Direction is similar to angle. It is specifically related to the orientation of a line in a coordinate system, which can be described as left, right, up, or down. A perception of three dimensions can be created in a two-dimensional figure by placing lines at carefully chosen angles relative to the other lines.

Intersecting lines create angles that provide a visual impact on visualizations, such as pie charts. The eyes see angles formed by the lines—not the lines. They notice wide angles and narrow angles based on the relative position of the lines. On a pie chart, the relative size of the slices is perceived based on the width of each angle—not on the wedge-shaped area that represents a slice.

Lines can distort, emphasize, or de-emphasize charts based on their length, angle, and direction. In Figure 7-16 (Chapter 7), two charts display the same data with identical scales. The difference is the aspect ratio (width-to-height proportion). In this graphic, the perceptual differences are significant. The top chart shows a steep curve indicating a rapid increase, while the bottom chart shows a slow and steady increase. The top chart shows significant peaks and troughs, while the bottom chart shows relatively minor declines.

Shapes

On a chart, the area and volume of shapes, such as a square or circle, communicate the relative size of the items that they represent. In visual perception, geometric shapes fill multiple roles, with examples such as circles, squares, triangles, and rectangles. Each shape can convey a specific meaning, such as a circle versus a square. The area of a shape communicates meaning—a large square indicates more than a smaller square. The shape aspect ratio may also denote something. A tall, thin, or thick bar can provide a different message. Therefore, care is needed to avoid miscommunication and misperception due to an object's shape or size.

Area

A key influence of shapes on visual perception is related to the area represented by a shape. Comparing an area conveys the relative size, importance, or magnitude of the items represented by the shapes. The actual area, however, does not always align with the perceived area. Careful use of shapes avoids miscues by understanding the differences between geometric areas and perceived areas.

3D Shapes and Volume

With three-dimensional figures, the potential for miscues can be magnified. For some viewers, the visual perception of relative size is based completely on the area of the shapes. However, for most viewers, it is based on the perception of the volume of the 3D object. Volume can distort visual comparisons based on the object's perceived depth. Perception of depth is based on the angles of the lines that connect the front and back of a shape.

Problem with Circles

It is challenging to judge the relative size of a circle. For example, what does half the size mean? For some it may mean half of the area, while for some it may mean half of the diameter. For judging relative size, comparing different types of shapes becomes even more difficult. For example, it is difficult to compare circles with squares, squares with triangles, and circles with triangles.

Impact of Shapes

Shapes are an integral part of visual design. A designer must be aware of the impact shapes can have and choose the appropriate shapes. Figure 8-2 shows the challenge of comparing shapes of different sizes. The first circle has the same height and width as the square, and the areas of the two shapes appear comparable. However, it is difficult to compare the area of the square with the second and third circles.

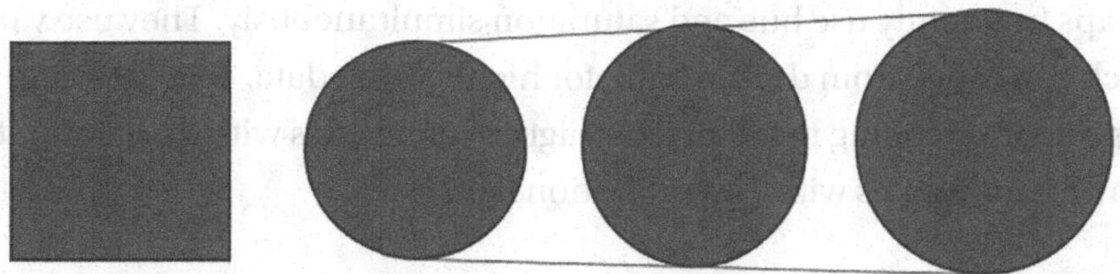

Figure 8-2. *Comparing shapes*

Color

Color is a very powerful visual cue that enables a large amount of data to be encoded in a small visual area. Hue, as well as saturation (intensity), influences the perceived importance of chart objects. The limitation of color is that many users are color blind, with 8% of males being red-green deficient.

Saturation is frequently used to encode categorical data, where each hue corresponds to a different group. Color is intense when saturation is high and it looks faded when saturation is low. More saturation indicates higher quantities, while less saturation indicates lower quantities.

Figure 8-3 uses color to encode categories, where all colors are saturated equally.

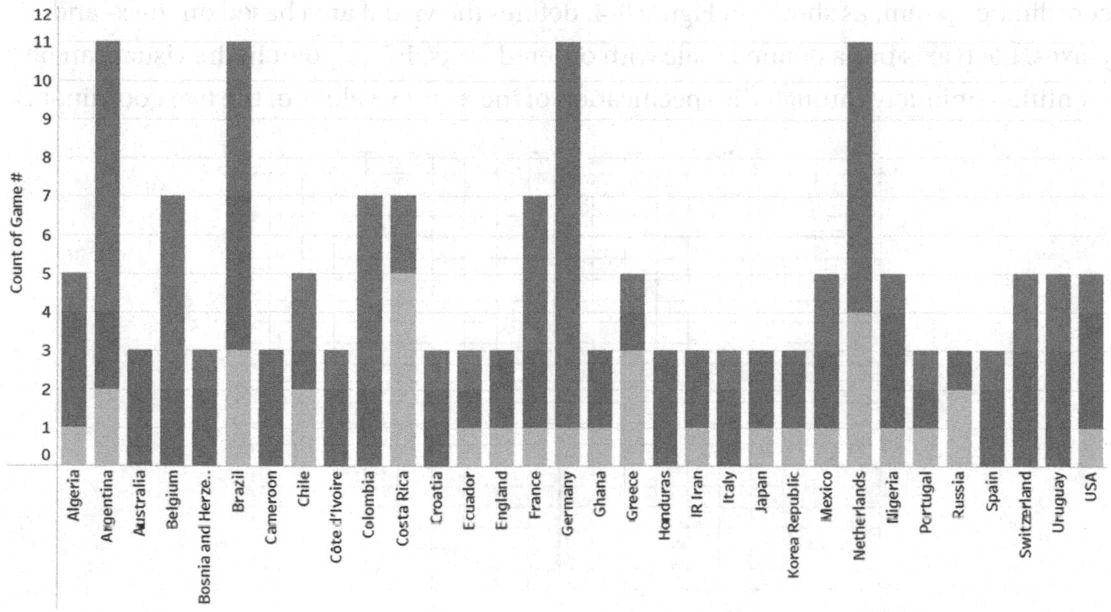

Figure 8-3. *Color encoding of categories (uses linear scale)*

Heat maps frequently use hue and saturation simultaneously. They use various colors, which can range from dark to light, for highlighting data. Together, hue and saturation provide encoding that indicates highest quantities with high-saturation dark hues and lowest quantities with low-saturation light hues.

Coordinate Systems

Coordinate systems enable quantitative values to be plotted to specific chart locations. They include Cartesian systems, polar systems, and geographic systems.

Cartesian Coordinates

A Cartesian coordinate system places data points on a two-dimensional grid. It is defined by an x-axis and y-axis, with examples being line charts and scatterplots. Encoding data includes placing visual objects meaningfully into a structured space. The quantitative values determine placement relative to x-y coordinates and latitude/longitude coordinates.

A coordinate system defines the structure of the structured space. A Cartesian coordinate system, as shown in Figure 8-4, defines the visual area based on the x- and y-axes. Each axis has a defined scale with ordered units. Every point in the visual can be identified uniquely through the specification of the x and y values of the two coordinates.

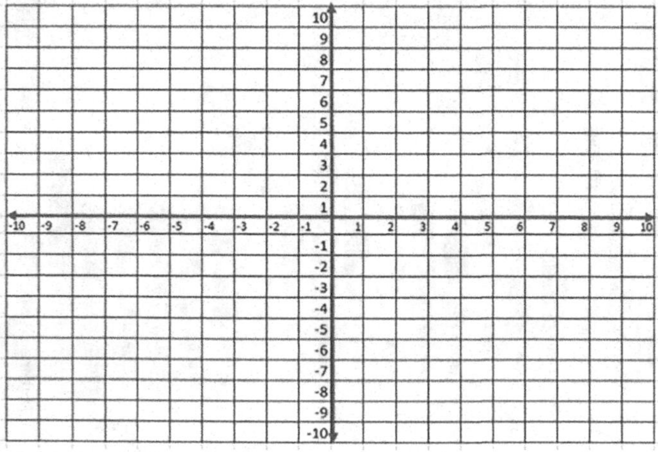

Figure 8-4. Cartesian coordinates

Polar Coordinates

A polar coordinate system places data points based on a circular system, where the center of the circle is the base point or zero point. It plots data points based on two coordinate values:

- Distance along a radius from the center point of the circle
- Angle of the radius in a range from 0 to 360 degrees

An example of a coordinate system is shown in Figure 8-5.

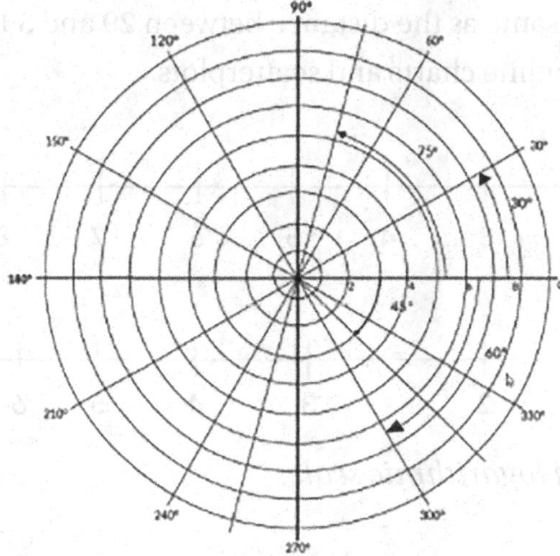

***Figure 8-5.** Polar coordinates*

Geographic Coordinates

A geographic coordinate system places data points on a geographic (two-dimensional) or geospatial (three-dimensional) map. The data point values determine placement in a coordinate system based on latitude and longitude. For three-dimensional plots, placement includes altitude. An example of a geographic coordinate system is a heat map.

Measurement Scales

Measurement scales define the association of data points with geographic coordinates in a data visualization area. They are the means by which data values are mapped to specific placement in the visualization area. They include linear scales, logarithmic scales, time scales, ordinal scales, and categorical scales.

Linear and Logarithmic Scales

Linear scales use equal divisions for equal values, as shown in Figure 8-6. The distance between 13 and 23 is the same as the distance between 29 and 39 or between 77 and 87. They are typically used for line charts and scatterplots.

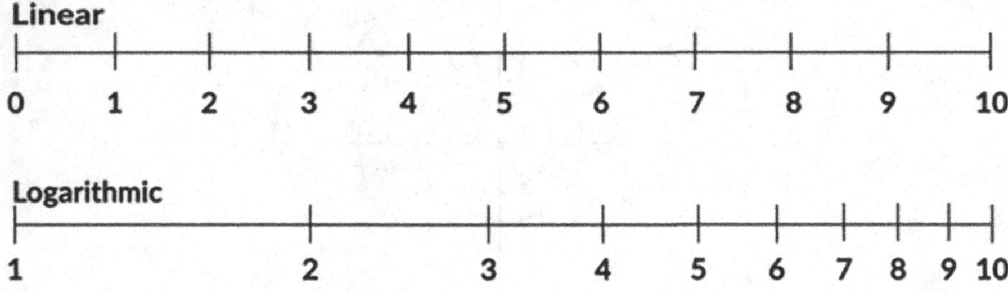

Figure 8-6. *Linear and logarithmic scales*

Logarithmic scales are similar to linear scales. However, their units are not of equal distance, as shown in Figure 8-6, because they are graduated using a multiplier. For example, the Richter scale for measuring the intensity of earthquakes is a base-10 logarithmic scale. An earthquake of magnitude 2 is ten times the strength of a 1 magnitude earthquake. An earthquake of magnitude 3 is ten times more powerful than magnitude 2 or 100 times more powerful than an earthquake of magnitude 1. This scale is used for charts and scatterplots when the range of values is very large. Figure 8-7 shows the difference between linear and logarithmic scales.

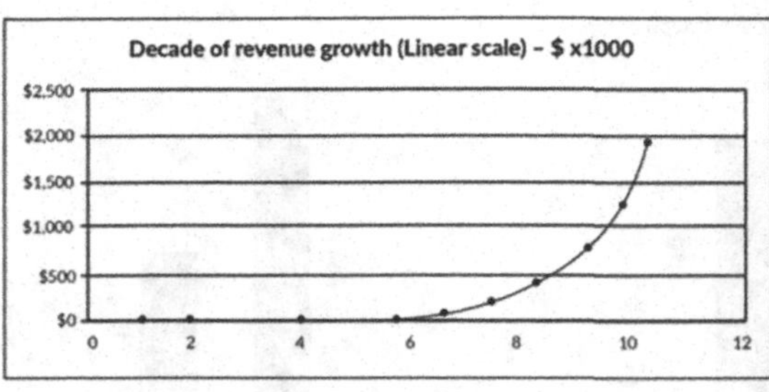

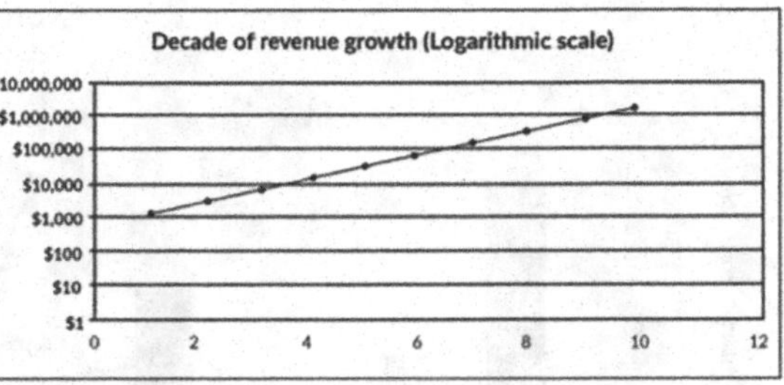

Figure 8-7. *Linear versus logarithmic scales comparison*

Ratio and Interval Scales

Linear and log scales indicate the distance between the scale division, while ratio and interval scales describe the characteristics of quantitative values attached to scale divisions, as shown in Figure 8-8 and Figure 8-9. Ratio scales are the most widely recognized measurement scales. Ratio variables never fall below zero. For example, height and weight measure from zero and above but never fall below it.

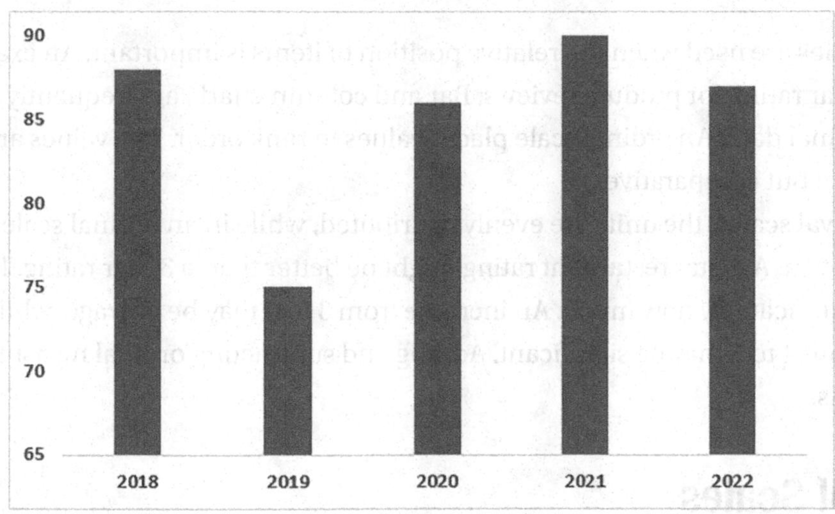

Figure 8-8. *Ratio scale*

The difference between ratio and interval scales is their ability to dip below zero. Interval scales hold no true zero and can represent values below zero.

Interval scales, shown in Figure 8-9, have no absolute zero point, which limits their ability to compare values. The scale units are distributed equally as with a ratio scale. However, zero point is arbitrary.

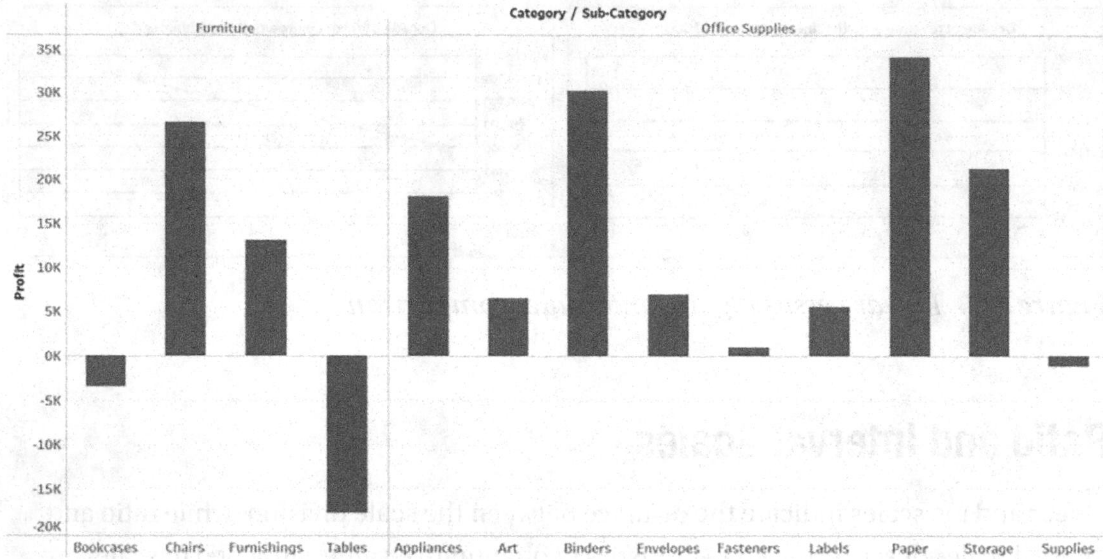

Figure 8-9. *Interval scale*

Ordinal Scales

Ordinal scales are used when the relative position of items is important. An example is the 1-5 star rating for product reviews. Bar and column charts are frequently used to display ordinal data. An ordinal scale places values in rank order. The values are non-proportional but comparative.

In interval scales, the units are evenly distributed, while in an ordinal scale they may be uneven. A 4-star restaurant rating might be better than a 3-star rating. However, it does not indicate by how much. An increase from 3 to 4 may be average, while an increase from 4 to 5 may be significant. Adding and subtracting ordinal measures is meaningless.

Nominal Scales

The nominal scale simply categorizes variables according to qualitative labels (or names). Examples include 1-mild, 2-moderate, and 3-severe, where the numbers serve only as "tags" or "labels" to identify or classify an object.

Percent Scales

Percent scales are a special case of linear scales, where the units are expressed as percentage values, as shown in Figure 8-10. Percent scales are used to display parts of a whole.

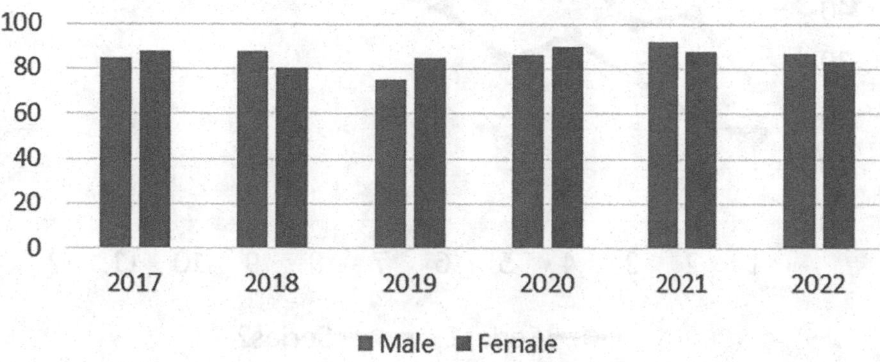

Figure 8-10. *Percent scale*

Time Scales

Time scales are a special case of a linear scale where the units are measures of elapsed time. The most common time series visualizations are line charts. Figure 8-11 shows a chart with a linear scale.

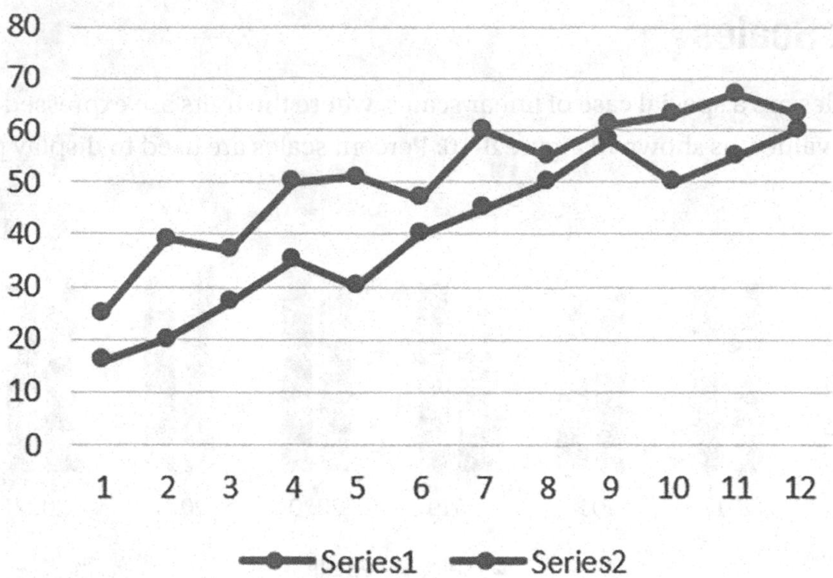

Figure 8-11. *Linear scale*

Visual Context

The fourth component of visual communication is visual context. Cues, coordinates, and scales can be combined to develop charts. Providing context on a chart enables a viewer to understand the scope and intended use of the visualization.

Explicit and Implicit Context

Context is important for understanding data, measures, and what the chart wants to convey. It can be provided by titles, legends, and annotations. In Figure 8-12, where no context has been provided, measures are categorized by cities, but it is not clear what is being measured. Figure 8-13 is a better chart as it displays implicit context.

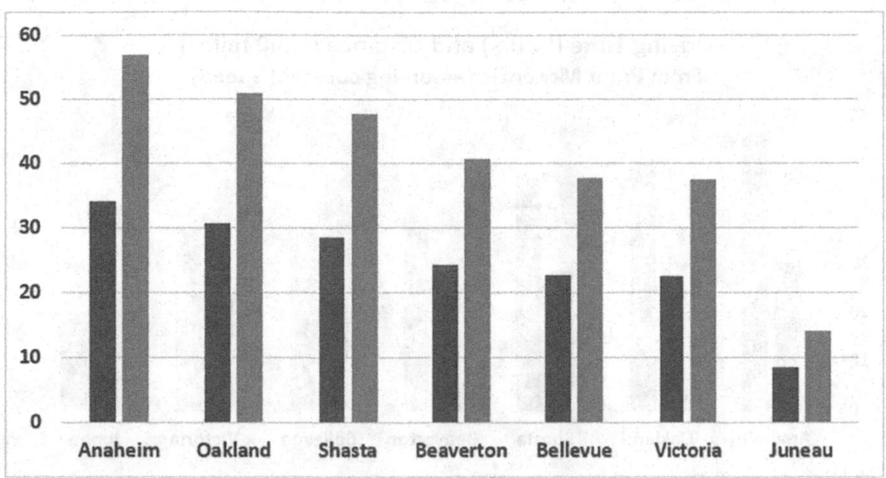

Figure 8-12. *Chart with no context*

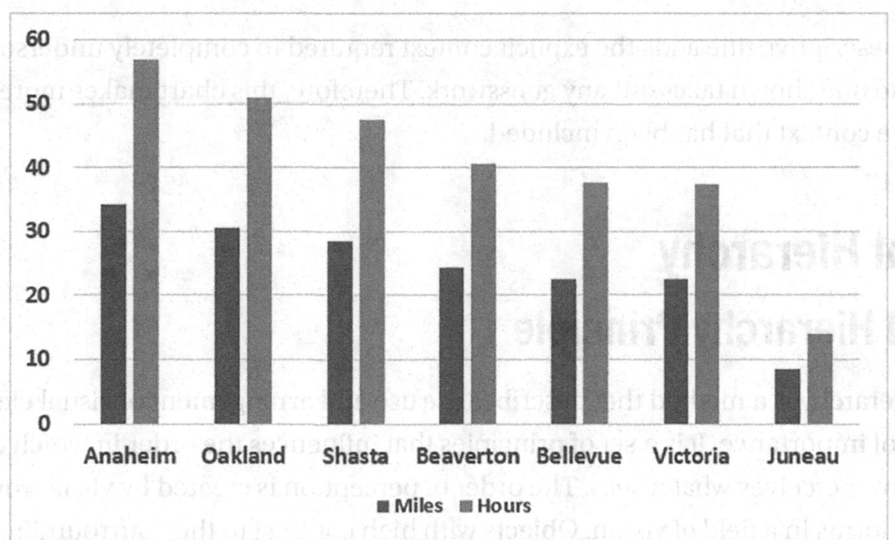

Figure 8-13. *Chart with implicit context*

The legend adds context and it also displays measures—miles and hours. The information supports some implications (implied context). It is reasonable to assume that the chart describes travel and the measures are travel distance and time. Figure 8-14 is more informative due to the explicit context that has been provided.

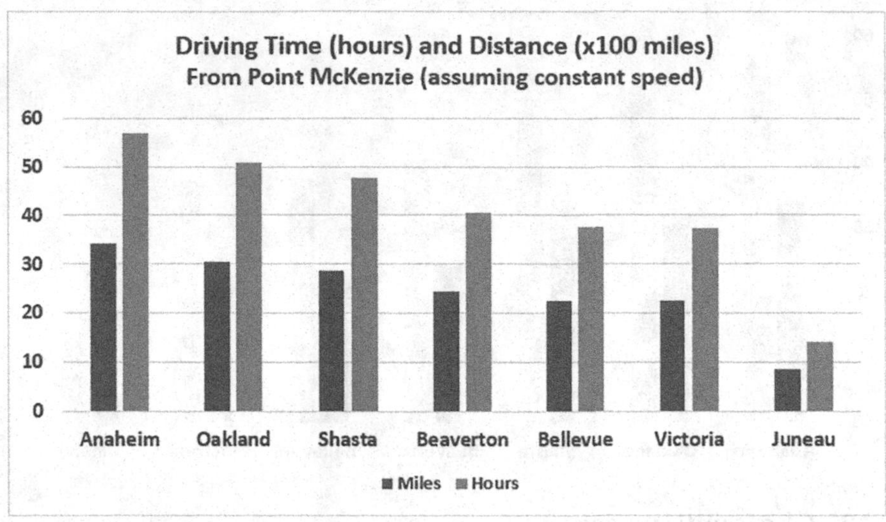

Figure 8-14. *Chart with explicit context*

The descriptive title adds the explicit context required to completely understand the chart. The title shown takes out any guesswork. Therefore, this chart makes more sense due to the context that has been included.

Visual Hierarchy

Visual Hierarchy Principle

Visual hierarchy is a method that describes the use and arrangement of visual elements in order of importance. It is a set of principles that influences the order in which the human eye perceives what it sees. The order of perception is created by visual contrast between forms in a field of vision. Objects with high contrast to their surroundings are recognized first and, therefore, viewed as most important.

As contrast decreases, the speed of recognition and perception of importance decline correspondingly. Visual hierarchy theory identifies four primary areas of contrast in order of visual impact:

- Color
- Size
- Alignment
- Character

Characteristics of Contrast Areas

The key characteristics of the four primary areas of contrast are:

- Color:
  - Involves both saturation and hue.
  - Bright hues are recognized before darker hues.
  - High saturation is recognized before low saturation.
- Size:
  - Can play an important role in the visual hierarchy area of contrast.
  - Bigger items are noticed before smaller items.
- Alignment:
  - It is less obvious but is often effective.
  - By skewing the position of an object, the eye is drawn to that object.
- Character:
  - Describes the degree to which a form is angular or curved.
  - Any contrast of character makes things stand out.
  - If nine out of ten data points are shown with circles and the 10th item is a square, the eyes get drawn to the square.

Summary

This chapter explored visual communication through four main components: visual cues, coordinate systems, measurement scales, and visual context. It explained visual cues, such as placement, lines, shapes, and color, which guide viewers' attention and perception accuracy. Also discussed were coordinate systems like Cartesian, polar, and geographic, which enable quantitative values to be plotted accurately. The chapter also described measurement scales, including linear, logarithmic, time, ordinal, and nominal. Visual context, including explicit and implicit, was also discussed along with visual hierarchy principles, emphasizing contrast in color, size, alignment, and character.

CHAPTER 9

Understanding and Using Charts

Visualizations for Dashboards

Variety of Visualizations

There are numerous visualization types available for display on a dashboard, which include bar charts, bullet charts, sparkline charts, scatterplots, heat maps, and treemaps. However, only a few work for the majority of needs.

The most common visualization types are:

- Heat map
- Scatterplot
- Line
- Slope graph
- Vertical bar
- Horizontal bar
- Stacked vertical bar
- Stacked horizontal bar
- Area
- Table
- Simple text

A. Khan, *Visual Analytics for Dashboards*, https://doi.org/10.1007/979-8-8688-0119-8_9

The two key attributes of a well-designed chart are that it should enable users to analyze the data and quickly extract relevant information. The most popular forms of this type of visualization include bar charts, pie charts, line charts, and area charts.

Tables interact with our verbal system, while charts interact with our visual system, which is faster at processing information. A well-designed chart typically gets the information across faster than a well-designed table. Simple text works when only a number or two need to be shared.

Plots and Maps

The most commonly used type of data visualization is the chart, which is of two types: plots and maps. The terms chart and graph are used interchangeably, though graphs form a subgroup of charts.

Plots are used to visualize the shape of data and observe patterns and relationships in data. Most plots place data points in a two-dimensional or three-dimensional visualization space, based on Cartesian coordinates. A two-dimensional plot has an x-axis and a y-axis, as shown in Figure 9-1, while a three-dimensional plot has x-, y-, and z-axes.

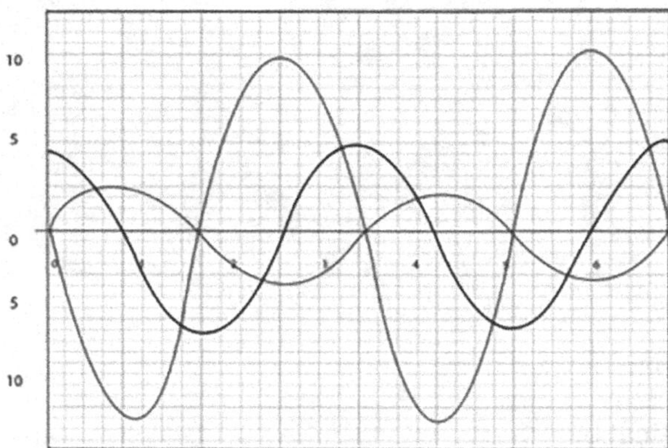

Figure 9-1. *Two-dimensional plot*

A map chart, shown in Figure 9-2, enables the visualization of spatial relationships in data by displaying data on a geographical map. Maps show the values of variables relative to their location in two-dimensional or three-dimensional space. They position data points using a geographic coordinate system.

Two-dimensional maps work with latitude and longitude. Three-dimensional maps extend the coordinate system by adding altitude. Maps are used when there is a need to view locations, location-based patterns, and location-based relationships in data.

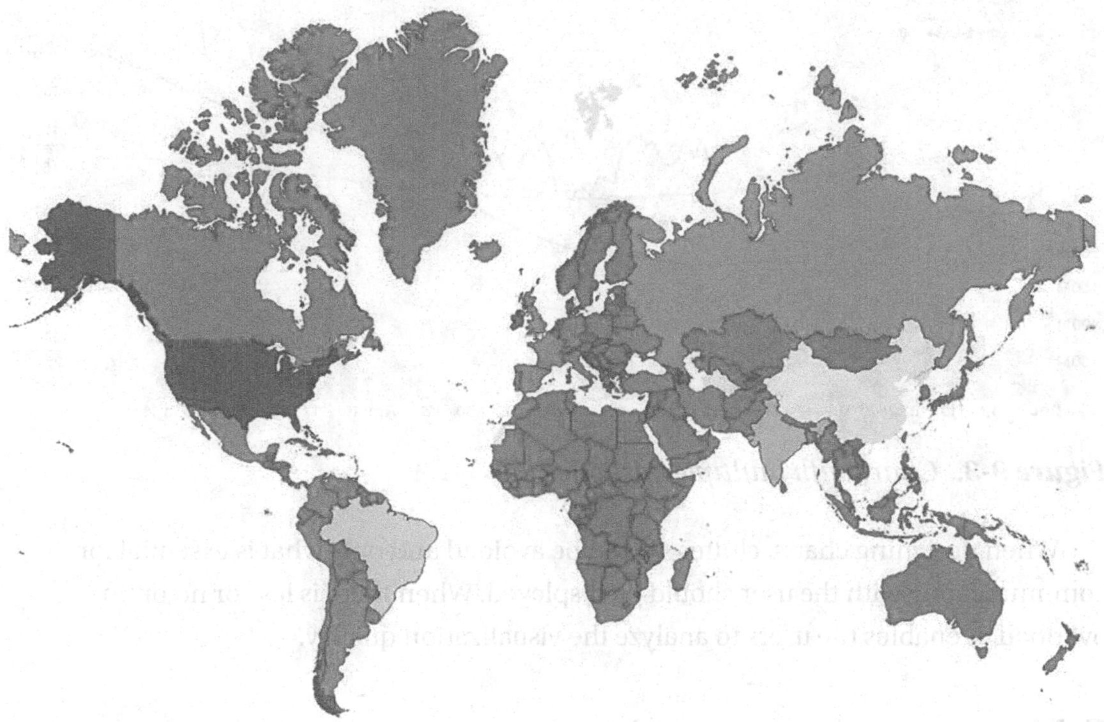

Figure 9-2. *Map chart*

Using Charts

Charts and graphs are diagrammatical illustrations of a dataset. Their differences, which are frequently not recognized, are mainly in the way the data is compiled and how it is represented. Graphs aim to focus on the data in question and how it trends. They have exact numerical figures shown on axes, usually organized on the left and bottom of the graph. They are most commonly used in analyses and situations that call for raw and exact data.

Charts are designed to show differences in a more aesthetically pleasing way. They are used primarily in business presentations. Both graphs and charts can have labels and legends.

Two different types of charts can be combined so that the analysis can be performed from different perspectives. For example, Figure 9-3 shows multiple variables being displayed in a single chart. The complex chart shows stock price performance over a six-month period combined with trading volume during the same time period.

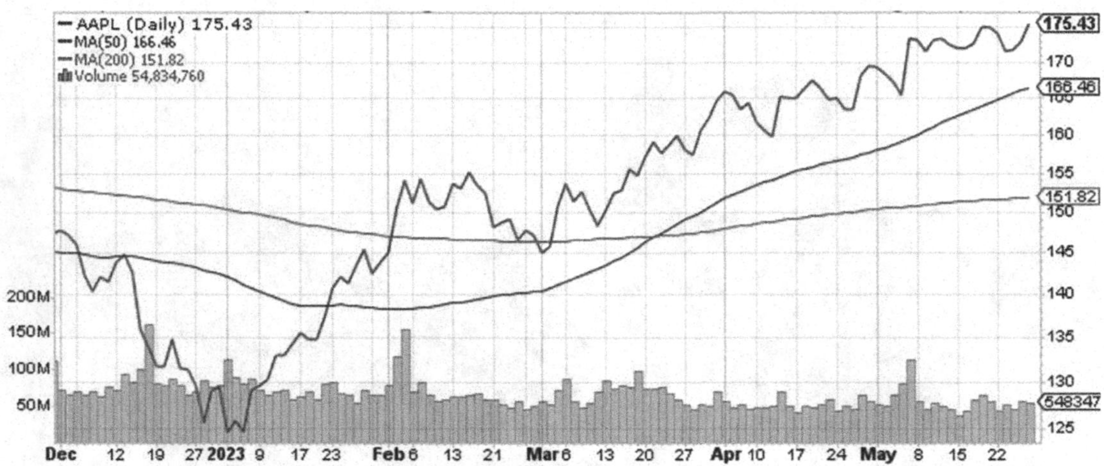

Figure 9-3. *Chart with multiple variables*

When designing charts, clutter should be avoided and only what is essential for communicating with the user should be displayed. When there is less or no brain overload, it enables the users to analyze the visualization quickly.

Tables

Visualization can also be done with tables, though they are not conceptualized when discussing data visualization. A table can visually present and arrange data, which is typically read as rows across and columns down. It can be quite useful when there is a need to view individual values, compare combinations of individual values, or view data that has precise values.

The concepts of cues and visual perception still apply when using tables for data presentation. The placement of data matters and includes both the sequence of rows and the order of columns. All of the following have a visual impact:

- Text color choices
- Presence and saturation of grid lines
- Fonts (bold, italic, and point size)

Table 9-1 displays a table where color is used to highlight specific cells.

***Table 9-1.** Table with color cues*

	Jan	Feb	Mar	Apr	May	Jun	Jul	Aug	Sep	Oct	Nov	Dec
Product 1	-36,921	635,741	143,354	66,795	-33,224	129386	27,592	112,796	657,971	730,404	77,743	414,120
Product 2	61,312	63,595	392,973	24,381	41,061	635,125	62,004	-48,018	55,758	338,111	-50,917	661,986
Product 3	308,666	7,365	36,738	340,578	364,345	117,856	45,710	171,634	377,492	284,672	56,780	81,305
Product 4	35,034	-21,363	57,595	90,700	24,089	393,579	343,614	75,233	11,961	1,155,289	87,884	162,944
Product 5	411,189	93,189	49,312	161,726	415,268	194,602	446,170	68,360	230,346	346,258	1,134,271	967,075

For effectiveness, table data should be prominent, which avoids heavy borders or shading that competes for attention. Light borders or simply white space should be used to set apart table elements. Borders can be used to improve table legibility. They can be pushed to the background by making them gray or they can be eliminated completely.

Chart Types

Column Chart

A vertical bar chart, also known as a column chart, displays data in rectangular bars—the longer the bar, the greater the value. A column chart plots the variable value vertically and the fixed dimension horizontally. It is used to compare values across categories and show change over a period of time. A bar or horizontal chart plots the variable value horizontally and the fixed dimension, such as time, vertically. The terms column chart and bar chart are frequently used interchangeably.

A column chart, like a line chart, can be single-series, two-series, or multiple-series, as shown in Figure 9-4. As more series are added, it becomes more difficult to focus on one at a time and perform meaningful analysis. Multiple-series bar charts should be used with caution. In charts that have more than one data series, visual grouping occurs due to spacing. Therefore, the relative order of categorization becomes important. Hence, the categorization hierarchy should be structured to make comparison easy.

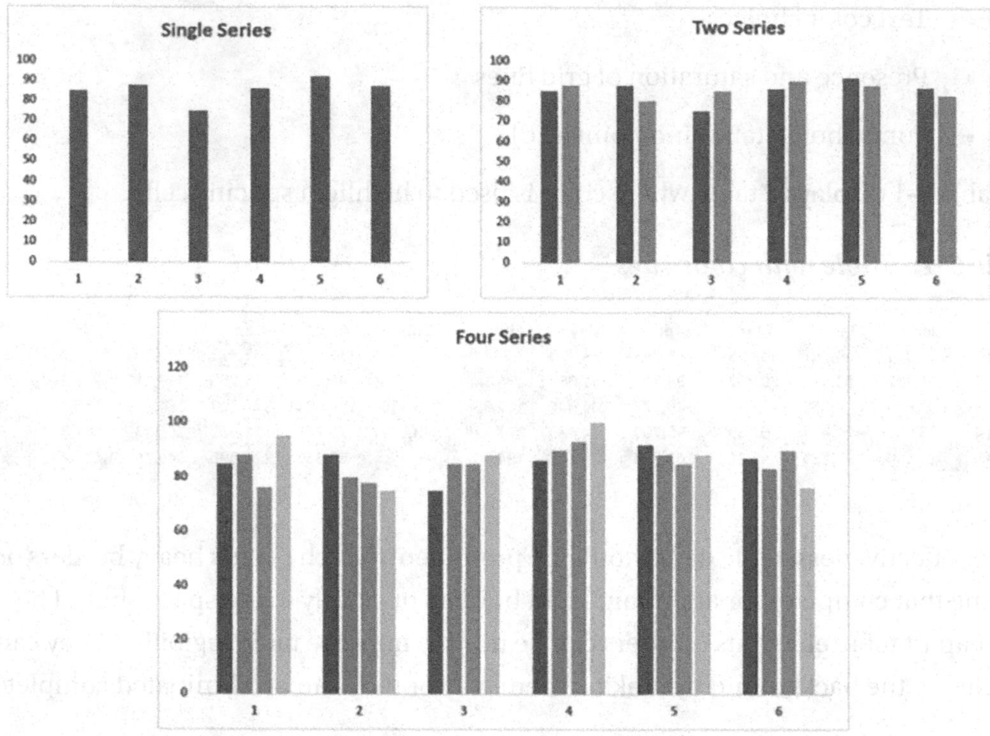

Figure 9-4. *Vertical bar charts (single, two, and multiple variables)*

Stacked Column Chart

There are limited use cases for stacked column (vertical) bar charts. They enable the comparison of totals across categories and also provide the sub-component pieces within a given category. Such charts can quickly become visually overwhelming, as it is difficult to compare sub-components across various categories. After moving above the bottom series (the one directly next to the x-axis), there no longer exists a consistent baseline that can be used for comparison.

Figure 9-5 shows an example of a stacked column chart, while Figure 9-6 shows a better way to represent the same stacked bar information.

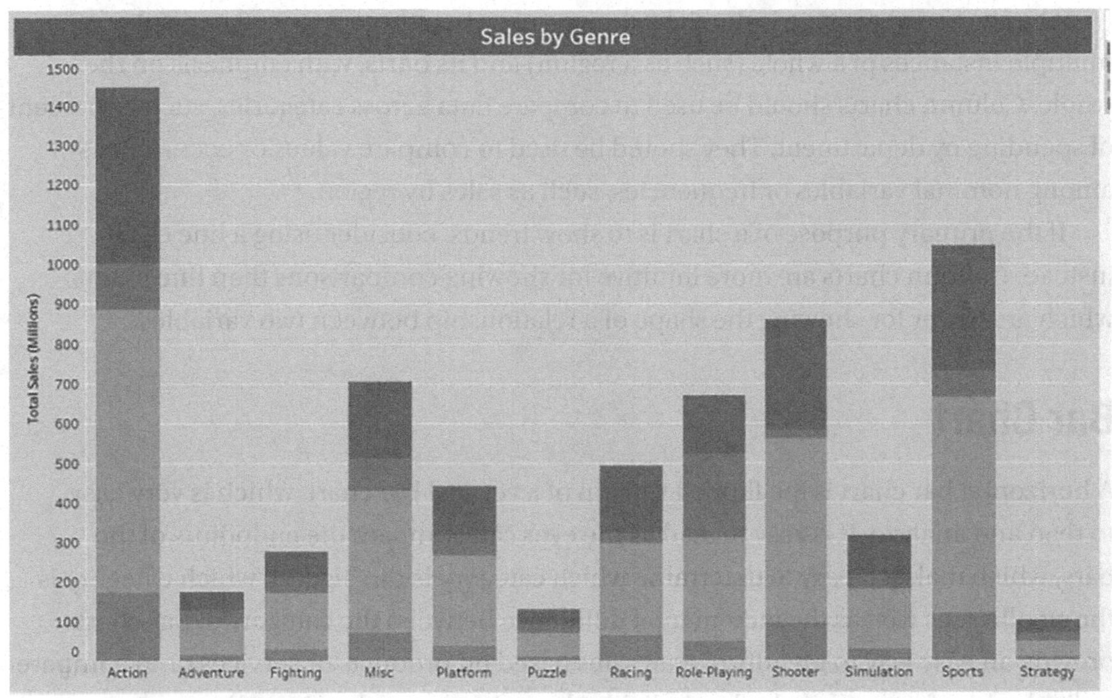

***Figure 9-5.** Stacked column chart*

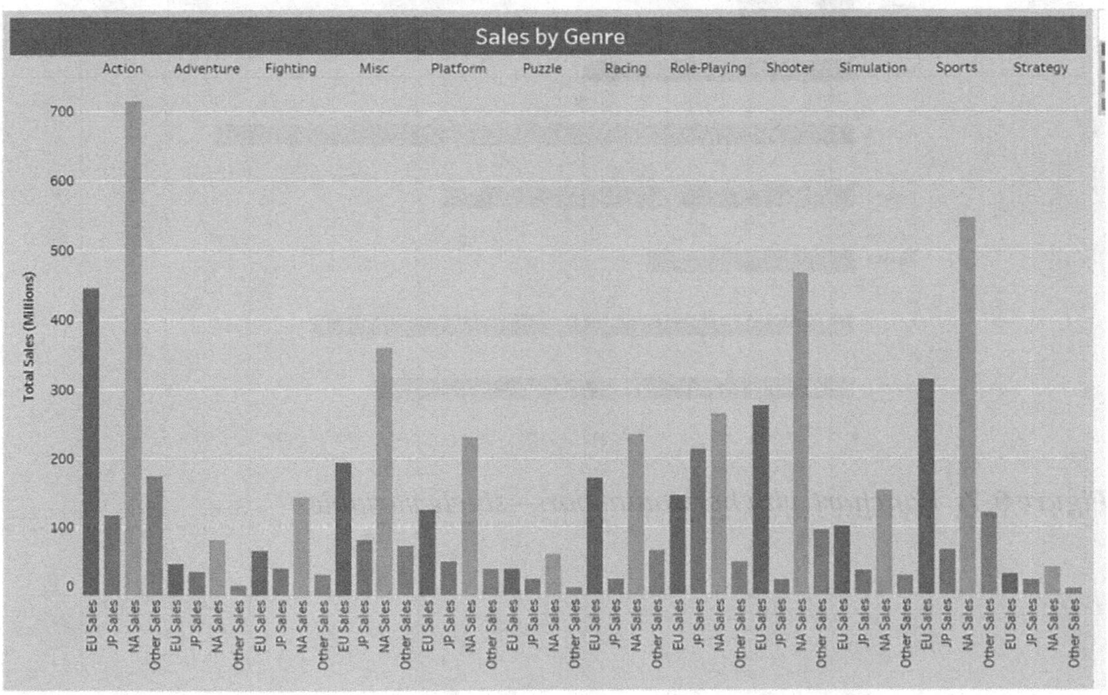

***Figure 9-6.** Better way to display stacked bar information*

The stacked chart type is the right choice only when there is a need to display multiple instances of a whole (such as a region) and its parts, with emphasis on the whole. Column charts should be used to compare data across categories, such as percent of spending by department. They should be used to compare values or occurrences among nominal variables or frequencies, such as sales by region.

If the primary purpose of a chart is to show trends, consider using a line chart instead. Column charts are more intuitive for showing comparisons than line charts, which are better for showing the shape of a relationship between two variables.

Bar Chart

A horizontal bar chart is the flipped version of a vertical bar chart, which is very easy to read and analyze. It is easy to read as the eyes can compare the endpoints of the bars, which makes it easy to determine which category is the biggest, which category is the smallest, as well as the incremental difference between the categories. Bar charts provide an easy way to identify overall sales based on products or services or to compare individual performance (whether individual, service, or product based).

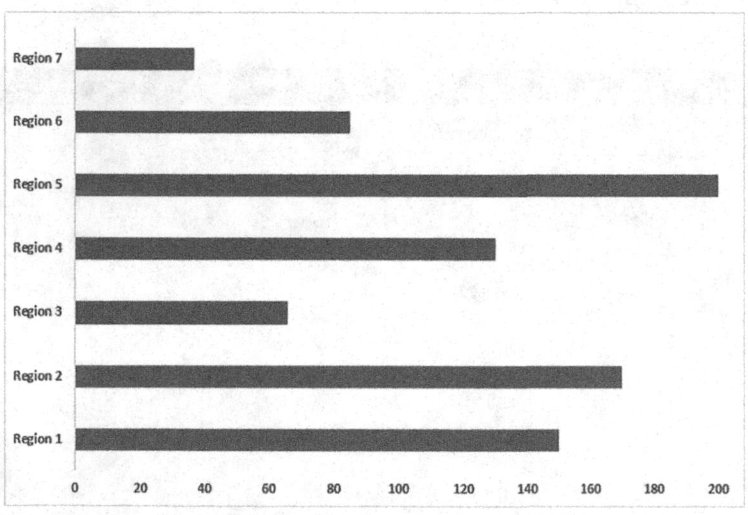

***Figure 9-7.** Bar chart with horizontal bars—single variable*

The purpose and usage of bar charts and column charts are the same. The choice of bars or columns is determined by visual style and formatting. Their usage in terms of providing comparisons of variables against one or more categories is similar.

Bar charts are especially useful if the category names are long because the text is written from left to right, which makes it easier to read. Due to the way information is processed, starting at the top left and then moving across the screen, the structure of the horizontal bar chart is such that the eyes hit the category names before the actual data. Before getting to the data, the user already knows what it represents and, therefore, does not need to dart back and forth between the data and the category names, as is the case with vertical bar charts.

Like the vertical bar chart, a horizontal bar chart can be single-series, two-series, or multiple-series, as shown in Figure 9-8.

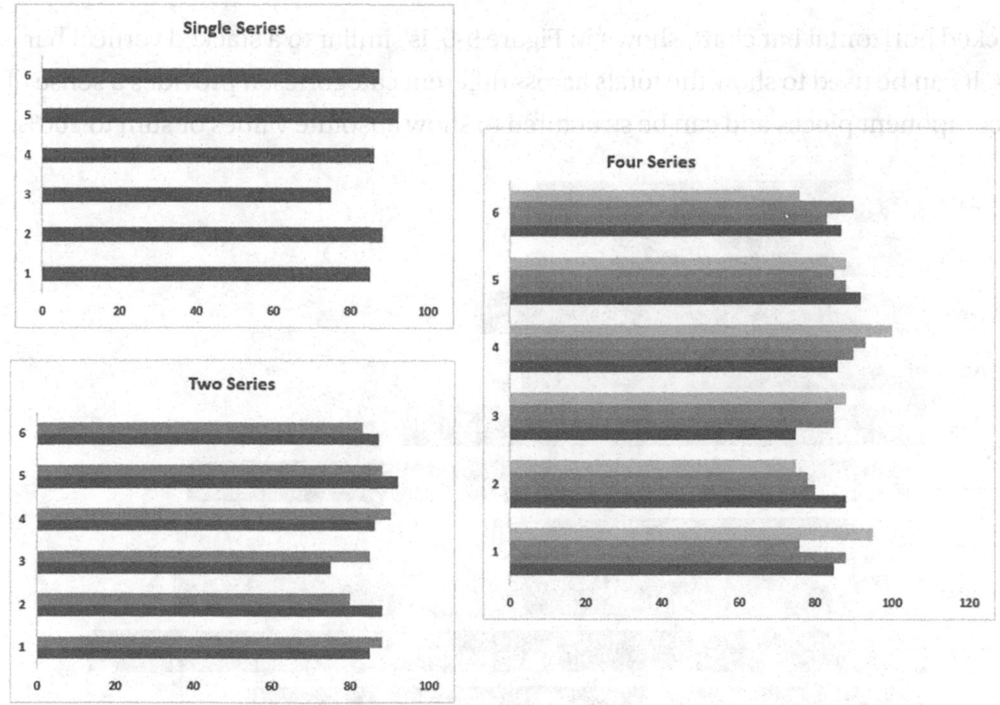

Figure 9-8. *Bar chart with horizontal bars*

Use a horizontal bar chart when there is a need to compare values or occurrences among nominal variables. Such a chart makes it easier to read labels. Horizontal bar charts are sometimes avoided because they are common, which is a mistake. They should be leveraged because most users are familiar with them and, therefore, the learning curve for the audience is shortened. Instead of trying to understand how to read the chart, users can use their brain power to determine which information can be extracted from the visual.

Bar charts should always have a zero baseline (where the x-axis crosses the y-axis) or a false visual comparison can result. Figure 7-15 (Chapter 7) shows how a non-zero scale can skew the view. The zero-baseline rule for bar charts does not apply to line charts. With line charts, the focus is on the relative position in space rather than the length from the baseline or axis, which allows having a non-zero baseline. In general, bars should be wider than the white space between the bars. Avoid making bars so wide that it appears areas are being compared instead of lengths.

Stacked Horizontal Bar Chart

A stacked horizontal bar chart, shown in Figure 9-9, is similar to a stacked vertical bar chart. It can be used to show the totals across different categories. It provides a sense of the sub-component pieces and can be structured to show absolute values or sum to 100%.

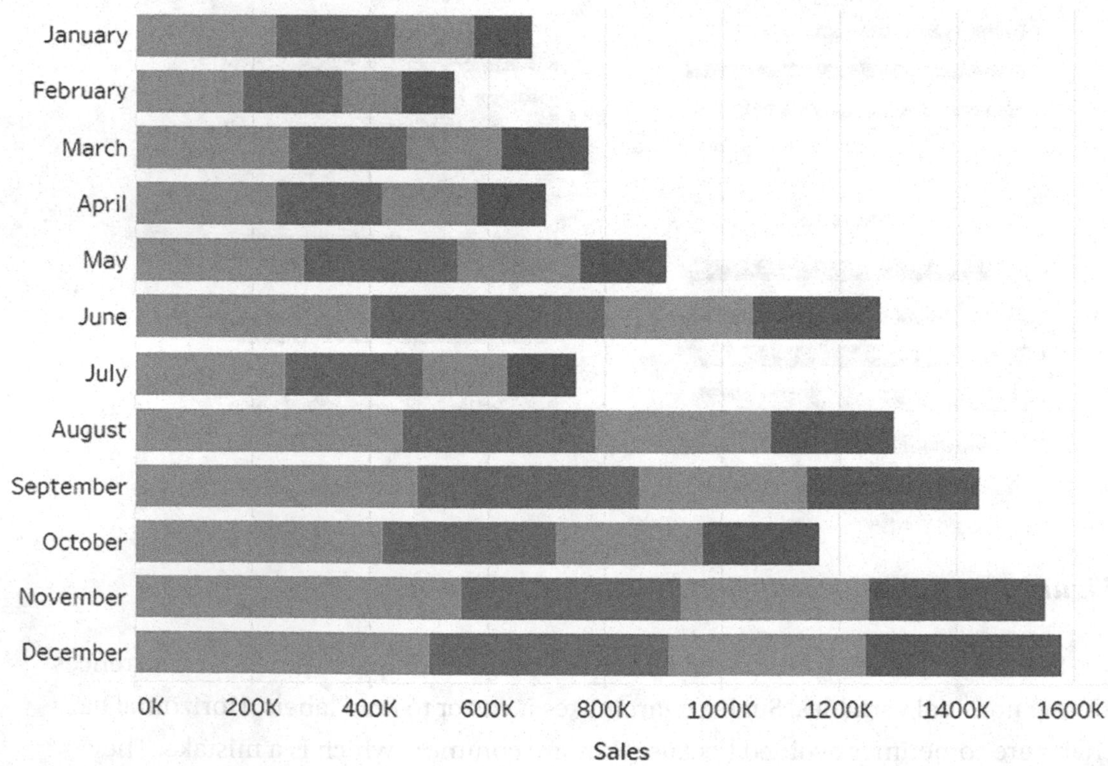

***Figure 9-9.** Stacked bar chart*

Pie Chart

The pie chart, shown in Figure 9-10, is one of the most popular types of visualization. It visually illustrates the relationship between a whole and its component parts, including the contribution that each component category makes to the whole. A pie chart can be used when there is a need to see proportional distribution, such as "How are sales distributed across the sales regions?" The problem with pie charts is that humans cannot effectively compare area or circular representations as well as the linear representation of data.

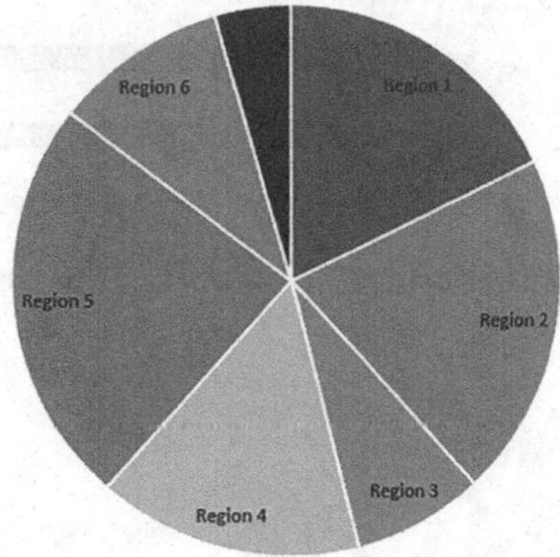

Figure 9-10. *Pie chart*

Pie charts tend to be overused, and most data visualization experts discourage their use. A pie chart can be useful when used carefully, especially when only a few categories are involved. The rule of thumb is to avoid using a pie chart when there are more than two to three categories. Some software tools enable pie charts to automatically group categories together if they fall under a certain percentile, which enables focusing on the larger numbers.

A bar chart is quicker and easier to process than a pie chart. In most cases, a bar chart can represent data better than a pie chart as shown in the Figure 9-11 example, where the bar chart is easier to analyze than the pie chart.

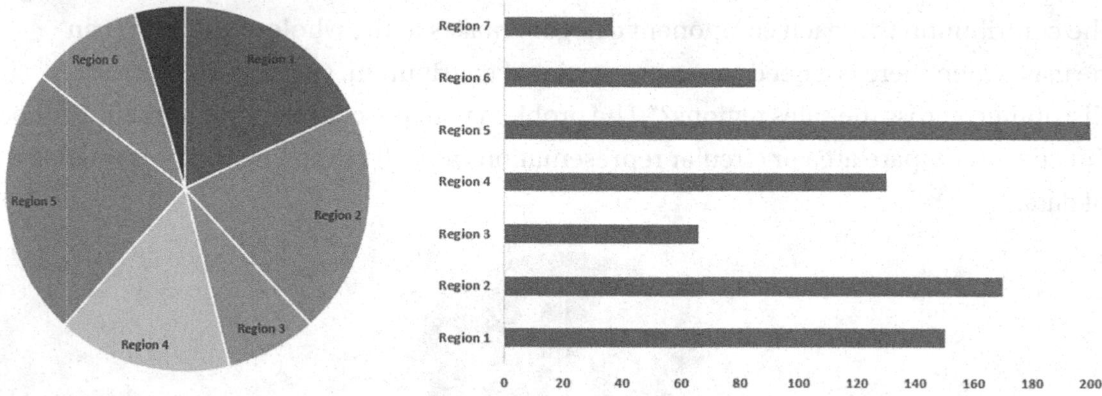

Figure 9-11. *Pie chart versus bar chart*

Donut Chart

A donut chart is similar to a pie chart, with the key difference being that the center of the donut is left open. An advantage of donut charts is that the area size distortion is less pronounced in the arc of a donut than in the slice of a pie. When used in an Infographic, the center of a donut chart is sometimes filled with an image for visual appeal.

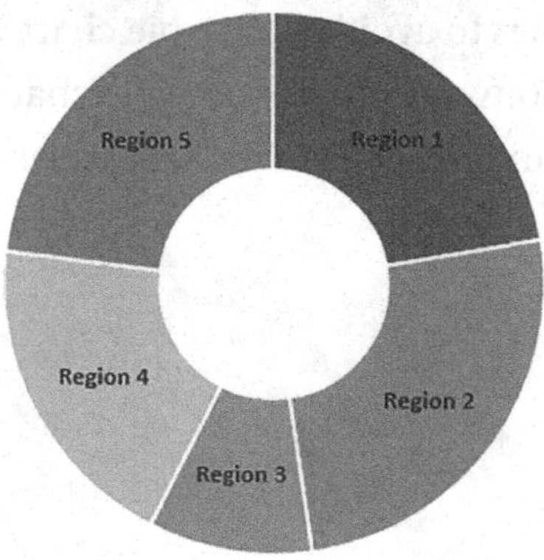

Figure 9-12. *Donut chart*

Line Chart

A line chart presents numerical data in a visual format that relates a dependent variable to an independent variable. It is commonly used to plot continuous data, frequently in some unit of time (days, months, quarters, or years), which is drawn as a set of lines between data points, as shown in Figure 9-13.

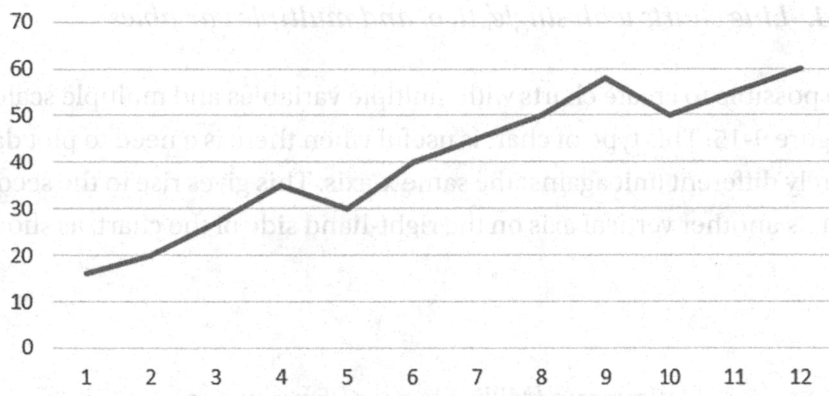

Figure 9-13. *Line chart showing a 12-month trend*

The purpose of line charts is to visually represent the shape of the relationship that exists between two variables (independent and dependent). Representing the data as a line implies that the x-axis represents a continuous variable. A line chart is most effective when there is a need to see values and trends over time. It can show if something is increasing, decreasing, or staying the same. For example, it can be used to determine how sales have increased or decreased over time.

When the x-axis represents discrete or categorical variables, a column chart is better. Time series analysis is the most common application of line charts in business analytics. It represents a business-dependent variable against time as the independent variable.

Line charts can show a single data series, two data series, or multiple data series, as shown in Figure 9-14.

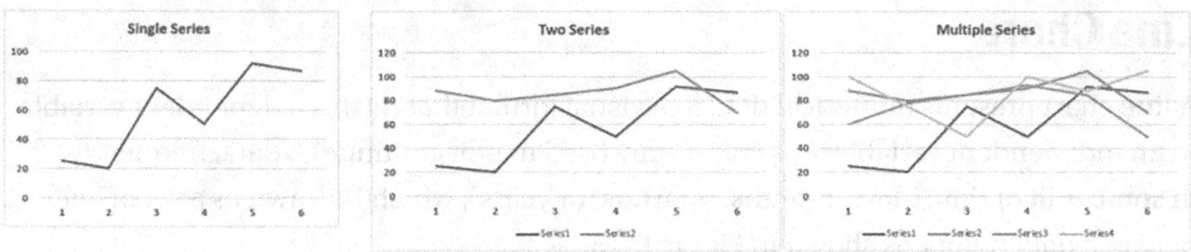

***Figure 9-14.** Line charts with single, two, and multiple variables*

It is also possible to create charts with multiple variables and multiple scales, as shown in Figure 9-15. This type of chart is useful when there is a need to plot data which is in an entirely different unit against the same x-axis. This gives rise to the secondary y-axis, which is another vertical axis on the right-hand side of the chart, as shown in Figure 9-15.

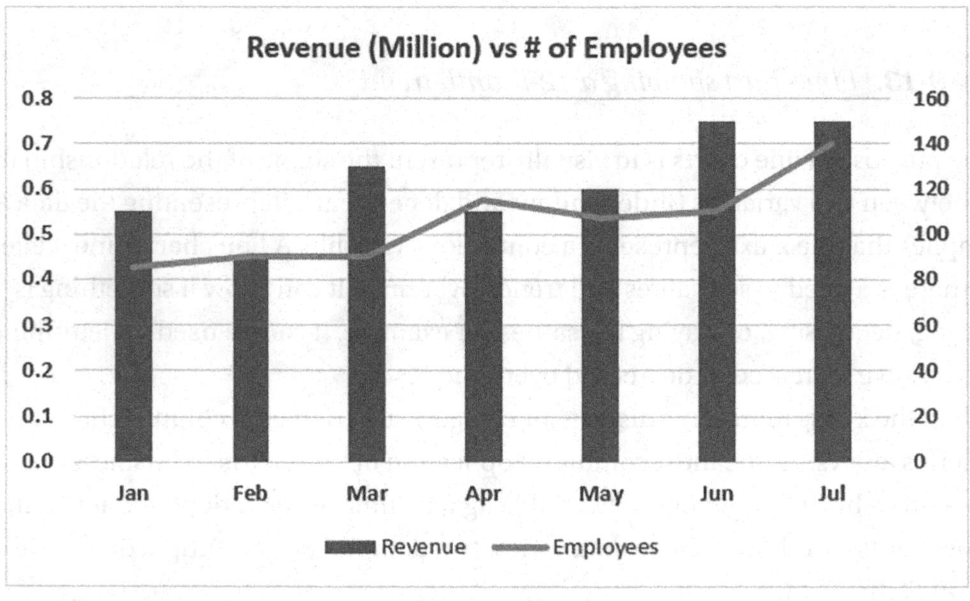

***Figure 9-15.** Multiple variables and secondary y-axis*

Figure 9-16 shows a bar chart and a line chart, which display the same information. However, the trend is much easier to determine in the line chart.

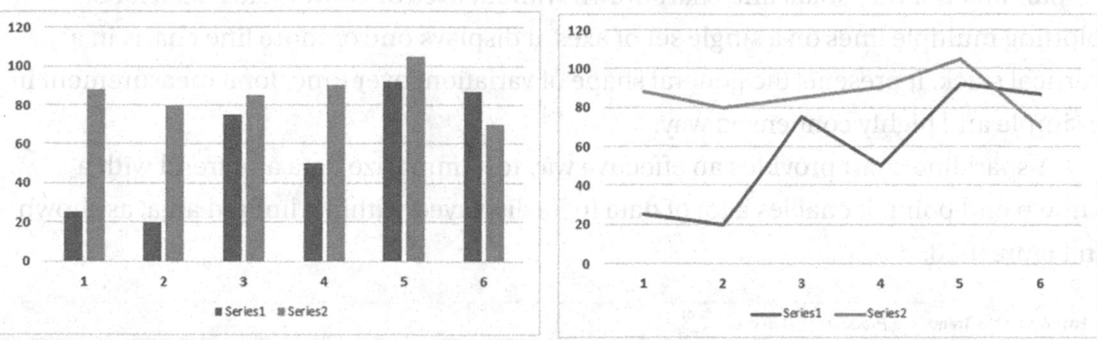

Figure 9-16. Bar chart versus line chart

Bars can be combined with trend lines to provide a different perspective, as shown in Figure 9-3.

Bullet Chart

The bullet chart is a variation of the bar chart. It serves as a replacement for dashboard gauges and meters, which typically display too little information, require too much space, and are cluttered with useless and distracting decoration or information.

A bullet chart compares a single quantitative measure (such as profit or revenue) against qualitative ranges (poor, satisfactory, good) and a target (such as the same measure a year ago). The dark horizontal middle bar represents the actual value, while the target is represented by the dark vertical line. The three colored or shaded bands represent ranges such as poor, satisfactory, and good or greater than the target value.

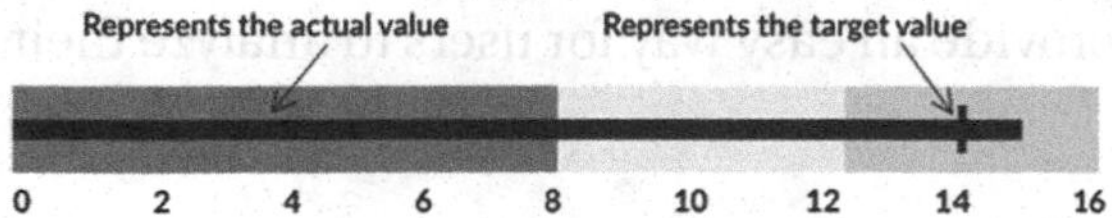

Figure 9-17. Bullet chart

Sparkline Chart

A sparkline is a very small line chart drawn without axes or coordinates. Instead of plotting multiple lines on a single set of axes, it displays one or more line charts in a vertical stack. It presents the general shape of variation, over time, for a measurement in a simple and highly condensed way.

A sparkline chart provides an effective way to summarize data as a trend with a known end point. It enables a lot of data to be displayed within a limited area, as shown in Figure 9-18.

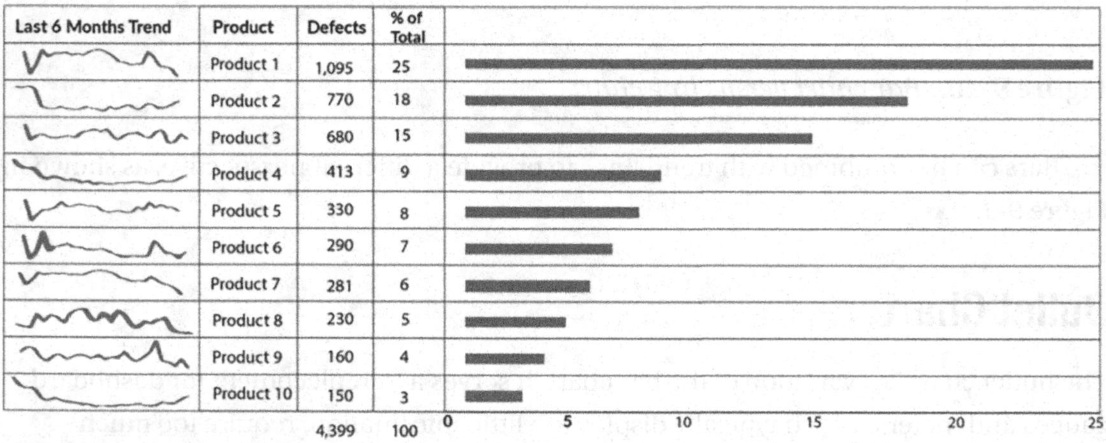

Last 6 Months Trend	Product	Defects	% of Total	
	Product 1	1,095	25	
	Product 2	770	18	
	Product 3	680	15	
	Product 4	413	9	
	Product 5	330	8	
	Product 6	290	7	
	Product 7	281	6	
	Product 8	230	5	
	Product 9	160	4	
	Product 10	150	3	
		4,399	100	0 5 10 15 20 25

Figure 9-18. *Sparkline chart*

Treemap

Treemaps are visualizations for displaying hierarchical data, which use nested rectangles, as shown in Figure 9-19. Each rectangle has an area proportional to the amount of data it represents. Treemaps provide an easy way for users to analyze their data at a glance.

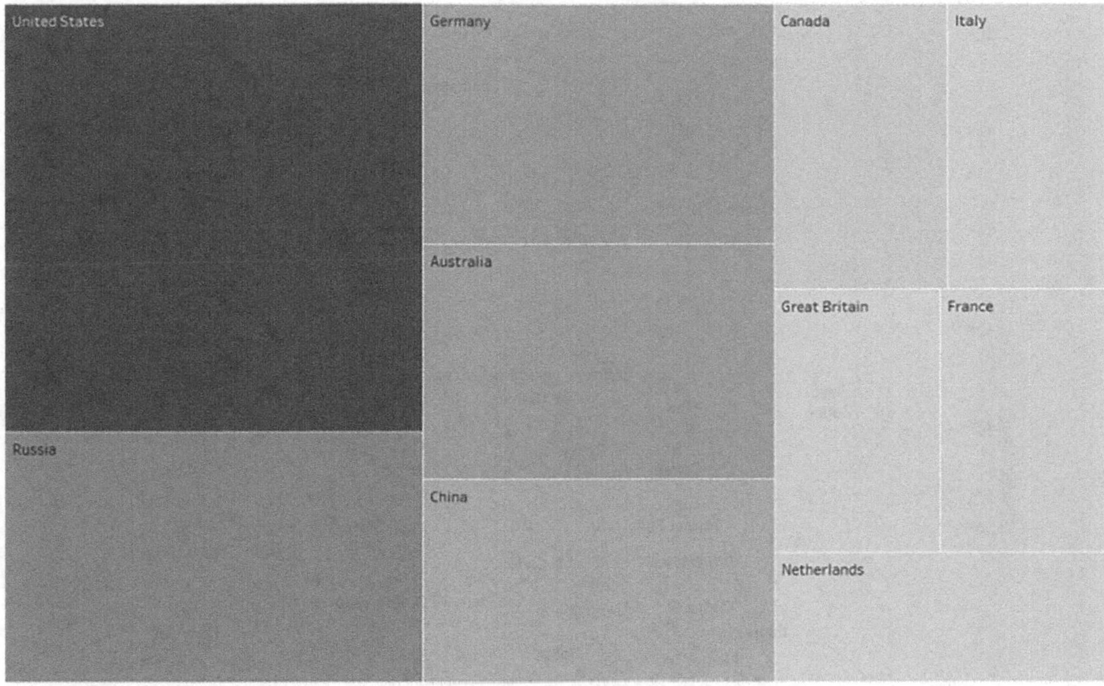

Figure 9-19. *Treemap: Olympic medal count by countries*

Scatterplot

A scatterplot, shown in Figure 9-20, is a chart whose objective is to display the relationship among variables—how much one variable is affected by another. In this type of chart, the values of two variables are plotted along two axes. The independent variable or attribute is plotted on the x-axis, while the dependent variable is plotted on the y-axis. The position of each mark, typically a dot, on the horizontal and vertical axis, indicates values for an individual data point. The pattern of the resulting points reveals any correlation that may exist.

Scatterplots are useful to understand dependencies inherent in the data, as well as the cause and effect, though not all correlations are causal. They are primarily used in the scientific world, though they are also used in business.

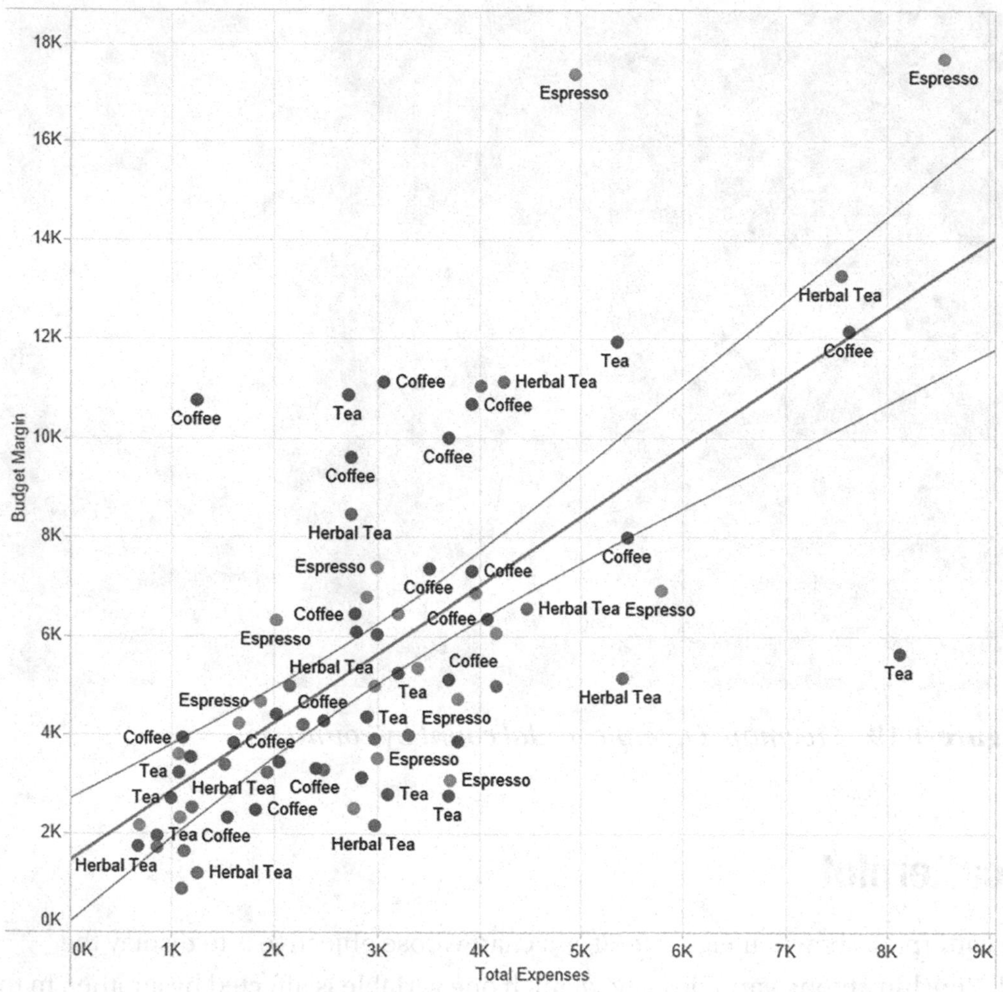

Figure 9-20. *Scatterplot*

Heat Map

A heat map shows the relationship between two variables, which are plotted on each axis. The data is represented graphically in which the values are represented as colors. By observing how cell colors change across each axis, patterns can be identified for one or both variables. The variation in color can be through hue or intensity, which provide visual clues regarding how the data is clustered or varies by location. A heat map can provide an efficient and comprehensive overview at a glance. It is an effective way to visualize large volumes of data in a small space.

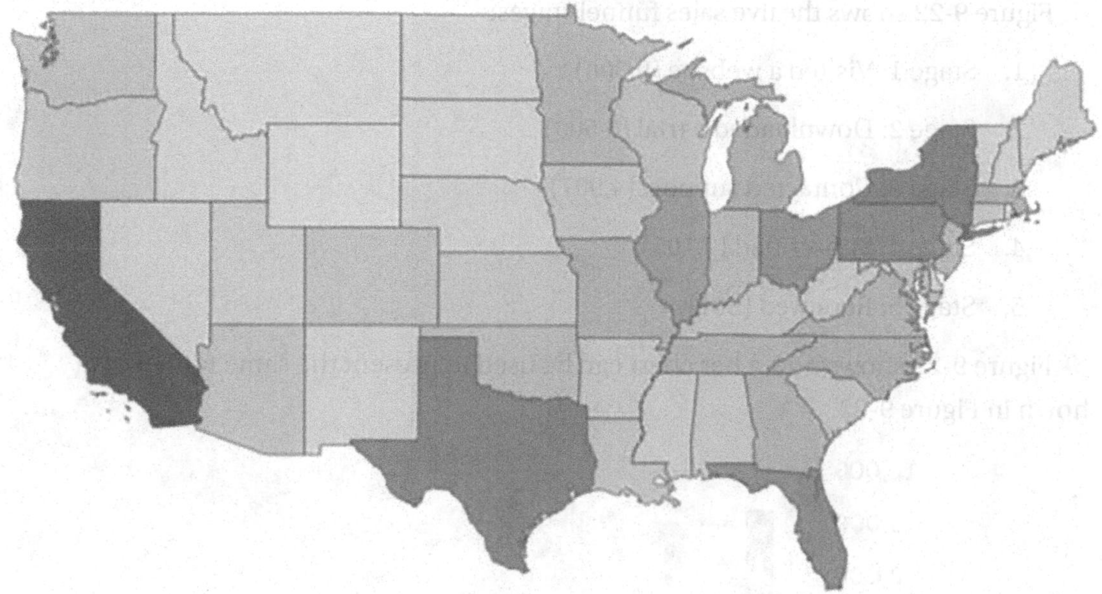

Figure 9-21. *Heat map*

Funnel Chart

A funnel chart shows values across multiple stages in a process, such as the flow of users through a business or sales process. For example, a funnel chart can be used to show the number of sales prospects at each stage.

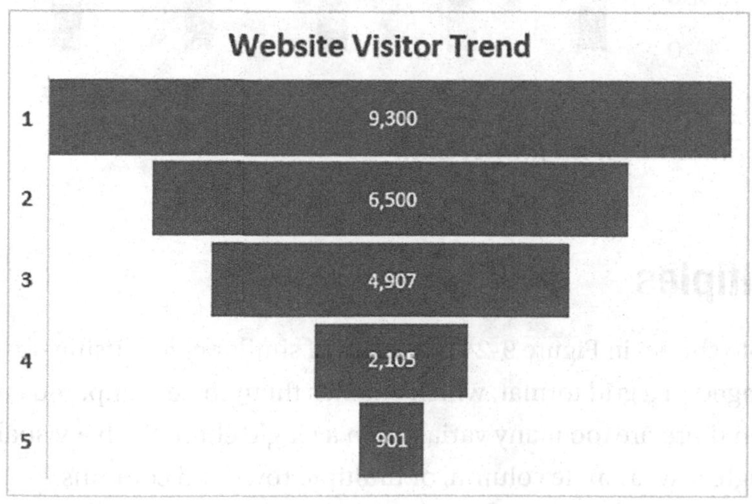

Figure 9-22. *Funnel chart*

Figure 9-22 shows the five sales funnel stages:

1. Stage 1: Visited a website (9,300)
2. Stage 2: Downloaded a trial (6,500)
3. Stage 3: Contacted support (4,907)
4. Stage 4: Subscribed (2,105)
5. Stage 5: Renewed (901)

Figure 9-23 shows how a bar chart can be used to present the same funnel data shown in Figure 9-22.

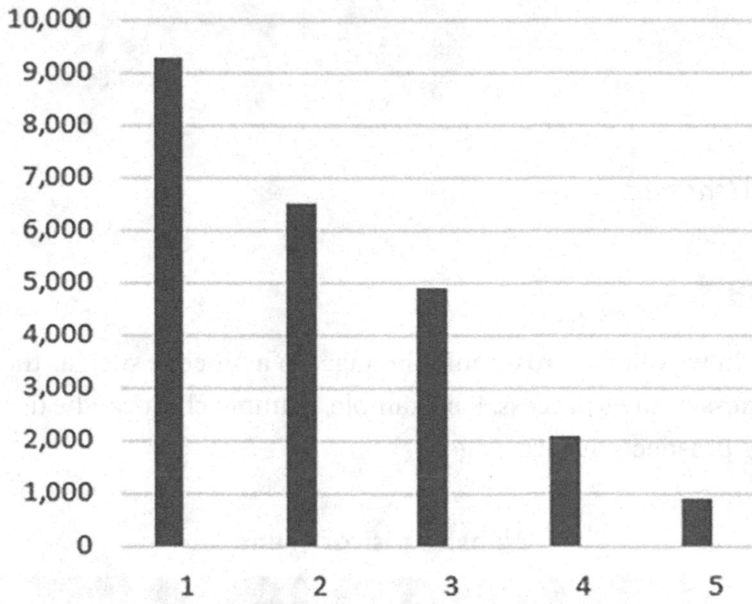

Figure 9-23. *Bar chart with funnel data*

Small Multiples

A small multiple shown in Figure 9-24 is a series of similar charts using the same scale and axes, arranged in a grid format, which enables them to be compared easily. They are useful when there are too many variables in a single chart. Such a visualization can consist of a single row, a single column, or multiple rows and columns.

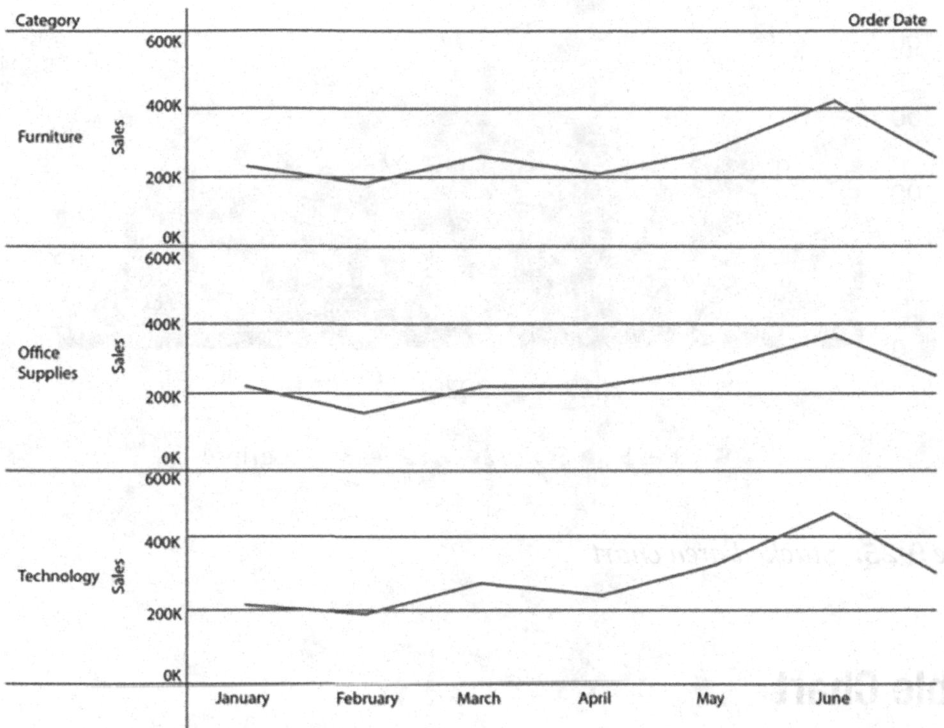

Figure 9-24. *Small multiples*

Area Chart

An area chart combines a line chart and a bar chart to show how quantities change over time. It shows how a group's numeric values change with the progression of a second variable. An area chart is similar to a line chart, where the data points are plotted and connected by lines. However, the difference is that an area chart is characterized by the addition of shading or color below the line, as shown in Figure 9-25.

In general, area charts should be avoided. The human eye does not do a great job of attributing quantitative value to two-dimensional space, which makes area charts harder to analyze compared to other types of charts.

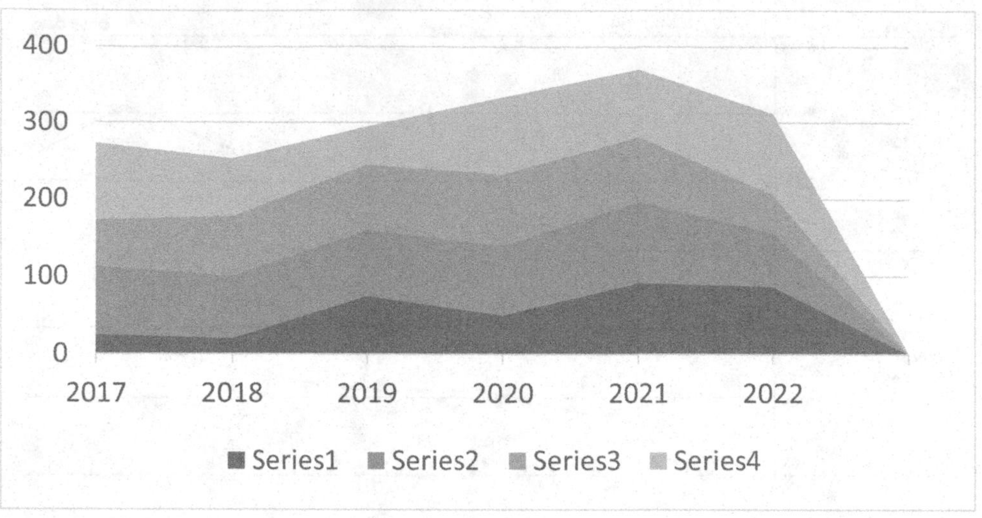

Figure 9-25. *Stacked area chart*

Bubble Chart

A bubble chart, which is an extension of the scatterplot, is used to visualize the relationships between three or more numeric variables. It represents three-dimensional relationships in a two-dimensional plane. This type of data visualization displays multiple bubbles or circles in a two-dimensional plot.

A two-dimensional grid is used to plot data points that relate two independent variables represented by values on the x- and y-axes. The third variable is represented by the size of the data point that is plotted. Instead of plotting a simple point, a small circle or bubble is plotted. The area of the bubble plotted is proportional to the value of the third variable being plotted along the x- and y-axes.

It is difficult to interpret this type of chart as it has high information density. However, it is powerful as it can convey multiple relationships in a two-dimensional plot. Bubble charts face visual perception issues associated with the relative size of circles. While it is easy to see which circles are larger or smaller than others, it is more difficult to accurately judge how much smaller or larger they are.

Figure 9-26 uses a grid similar to that of a line chart or column chart. It relates profit, cost, and the probability of success. The data points are plotted as bubbles representing the value of the third variable.

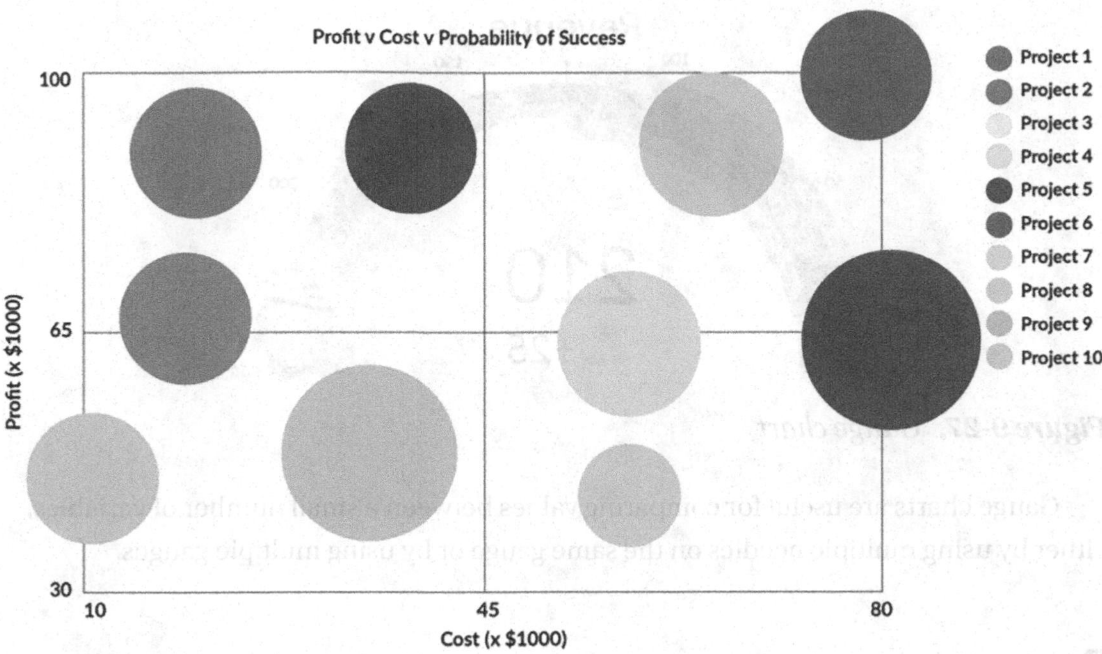

Figure 9-26. *Bubble chart*

Gauges

A gauge chart, also known as a dial or speedometer chart, provides an easy and quick way to determine how a metric is performing against a target. For this purpose, as shown in Figure 9-27, it uses a needle as a reading on a dial. The needle value is read against the colored data range or chart axis. A gauge chart consists of a gauge axis, which contains the data range, color ranges, interval markers, as well as a needle.

Figure 9-27. *Gauge chart*

Gauge charts are useful for comparing values between a small number of variables, either by using multiple needles on the same gauge or by using multiple gauges.

Summary

This chapter delved into the world of data visualization, identifying the most common visualization types, and explained how tables interact with our verbal system, while charts interact with our visual system, which is faster at processing information. It explained the two key attributes of a well-designed chart and also covered a wide array of visualization types in depth, including bar chart, stacked bar chart, line chart, pie chart, heat map, bullet chart, sparkline, scatterplot, area chart, treemap, and tables.

CHAPTER 10

Selecting Charts

Visualization Requirements

Purpose

A chart's primary goal is to describe and communicate the shapes that represent properties of and relationships among quantitative variables. Each property or relationship type corresponds with data visualization functions. By understanding the key elements of different visualization types, a designer can select the appropriate charting methods.

What Needs to Be Visualized

Table 10-1 is a guide that describes the scenarios in which various visualization techniques can be used.

***Table 10-1.** Guide for chart usage*

Visualization technique	Scenario
Whole and part	Show component breakdown of a quantitative variable (whole) into contribution levels of components (parts)
Simple comparison	Show quantitative values of a single variable for visual comparison
Multiple comparisons	Show quantitative values of multiple variables for visual comparison
Trends	Show how quantitative variable changes compared to a continuous, categorical variable; most common categorical variable for comparison is time
Frequencies	Show how a set of values is distributed across its full range, from lowest to highest, and how often each value occurs in the dataset

(*continued*)

A. Khan, *Visual Analytics for Dashboards*, https://doi.org/10.1007/979-8-8688-0119-8_10

Table 10-1. *(continued)*

Visualization technique	Scenario
Correlations	Show if the quantitative values from a paired set vary in relation to each other, as well as the direction of variance (positive or negative) and strength of dependency (strong or weak)
Spatial relationships	Show how quantitative variables are associated with location-based variables

What Drives the Selection

Selecting the Type of Chart to Use

Table 10-2 shows the situations in which specific chart types can be used.

Table 10-2. *Guide for choosing appropriate charts*

Chart type	When to use
Bar chart	Use when comparing data across categories, such as cars sold in different states or spending percentage by department
Line chart	Use for viewing trends over time, such as year-to-date stock price or sales channel revenue by month
Pie chart	Use for showing proportions, such as sales by region
Map chart	Use for displaying geocoded data, such as accidents by zip code or crime rate by state
Scatterplot	Use for determining the relationship between different variables, such as the likelihood of having Covid, for males and females, in different age groups
Bubble chart	Use for showing the concentration of data along two axes, such as sales by geography and product

(continued)

Table 10-2. *(continued)*

Chart type	When to use
Histogram	Use for understanding the distribution of data, such as the frequency of a product's failure
Bullet chart	Use for analyzing the performance of a metric against a goal, such as actual costs versus budget
Heat map	Use for showing the relationship between two variables and how data is clustered or varies by location, such as rainfall across states

Best Fit

Table 10-3 displays the mapping of different chart types, which can also help identify the appropriate chart for specific requirements.

Table 10-3. *Mapping chart types*

Chart type	Bar chart	Column chart	Line chart	Area chart	Scatter plot	Pie chart	Donut chart	Bubble chart
Trends	Yes	Yes	Yes	Maybe	Maybe	No	No	Maybe
Correlation	No	No	Yes	Maybe	Yes	No	No	No
Frequencies	Yes	Yes	Yes	Maybe	Yes	No	No	No
Simple comparison	Yes	Yes	Maybe	Maybe	Maybe	Yes	Yes	Maybe
Multiple comparison	Yes	Yes	Maybe	No	No	No	Maybe	Yes
Spatial relationship	No	No	No	Maybe	No	Maybe	Maybe	Maybe
Whole and parts	Maybe	Maybe	No	Maybe	No	Yes	Yes	No

Finding the Appropriate Chart

Deciding which chart to use to communicate the desired message is a challenging task because too many choices are available. Therefore, the use and application of each type of chart should be understood to ensure that the appropriate chart is selected for the data being analyzed.

Mapping Categories

There are four categories that indicate the type of visualization which should be used for a particular application. These are comparisons, proportions, relationships, and patterns.

- Comparisons:
  - Illustrate "What differences" between things
- Proportions:
  - Show "Which parts" comprise a whole
- Relationships:
  - Show "Which dependencies" exist among subjects or variables
- Patterns:
  - Distribution patterns: Show at "Which frequencies" data values or conditions occur
  - Locations patterns: Show "Which places" are associated with specific data values or conditions
  - Probabilities patterns: Show "What are the chances" that a specific value or condition will occur
  - Behavior-over-time patterns: Show "Which trends" are observed in the data

Comparisons

This category of charts can be used to determine similarities and differences between values. The common ways to show comparisons are bar charts and column charts, as shown in Figure 10-1.

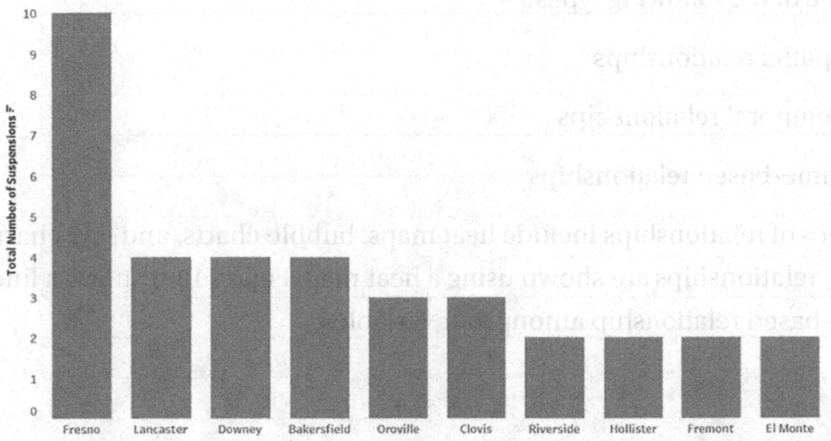

Figure 10-1. *Comparison*

Proportions

These visualization methods use area to show similarities and differences between values or as compared to a whole. The common visualizations in this category include pie charts, shown in Figure 10-2, and area charts.

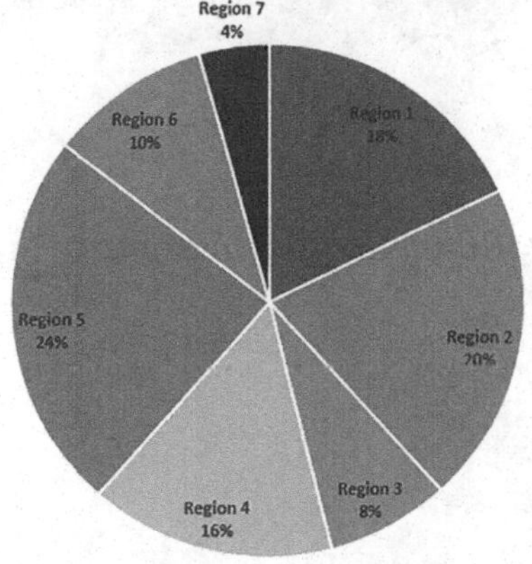

Figure 10-2. *Proportions displayed by a pie chart*

Relationships

These charts are used to visualize associations that exist in data, including relationships among entities and relationships among variables. These associations can exist at a subject level (i.e., associations among entities) or as relationships among variables, which can be of the following types:

- Spatial relationships
- Temporal relationships
- Time-based relationships

Examples of relationships include heat maps, bubble charts, and line charts. In Figure 10-3, relationships are shown using a heat map. Figure 10-4 shows a line chart with a time-based relationship among four variables.

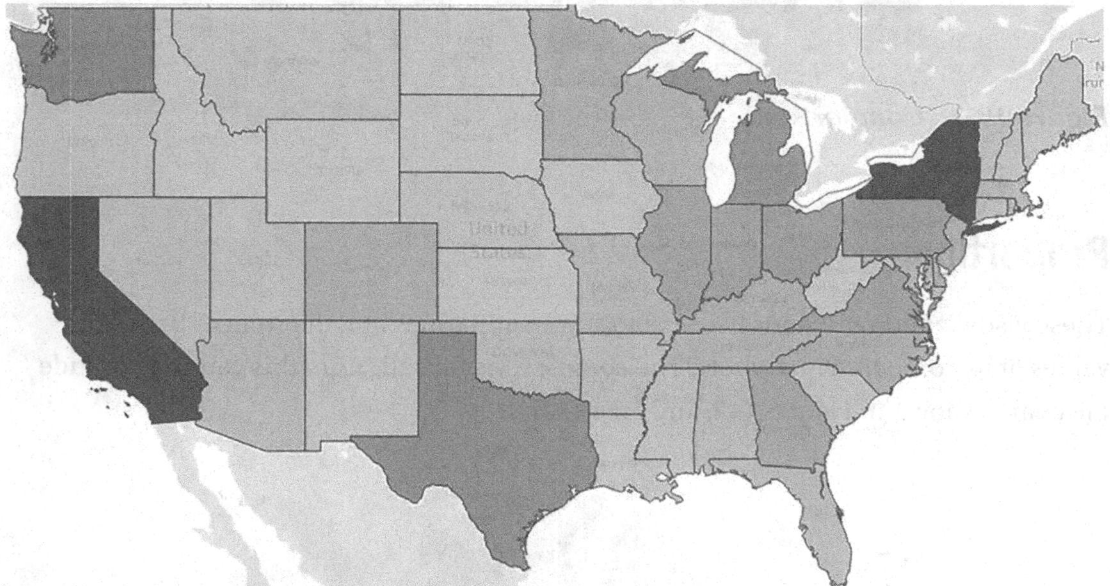

***Figure 10-3.** Relationships shown using a heat map*

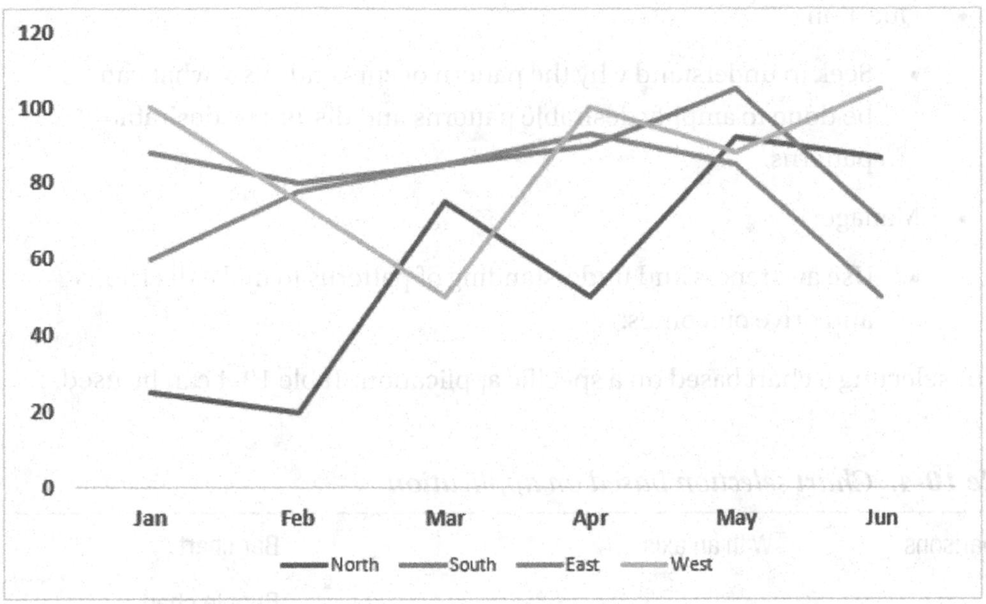

***Figure 10-4.** Line chart showing time-based relationship among four variables*

Patterns

Most of these types work well to show point-in-time patterns. However, they are not suited to the visualization of time series and continuous data patterns. These charts can be used to see similarities and differences between values.

Every chart type can reveal patterns that provide meaning to data. However, it is not possible to see patterns in table rows and columns. Charts and graphs enable patterns to stand out in ways that make data actionable. Patterns enable users to:

- Observe:
    - Be aware of what is happening.
- Associate:
    - Seek to understand pattern-based relationships with other variables.

- Question:
    - Seek to understand why the pattern occurs and, also, what can be done to amplify desirable patterns and disrupt undesirable patterns.
- Manage:
    - Use awareness and understanding of patterns to make decisions and drive outcomes.

For selecting a chart based on a specific application, Table 10-4 can be used.

Table 10-4. *Chart selection based on application*

Comparisons	With an axis	Bar chart
		Bubble chart
		Bullet chart
		Line chart
		Stacked area chart
		Stacked bar chart
	Without an axis	Choropleth chart
		Donut chart
		Heat map
		Pie chart
		Tree map

(*continued*)

Table 10-4. *(continued)*

Proportions	Proportions between values	Bubble chart
		Stacked bar chart
		Dot matrix chart
	Proportions in parts to a whole relationship	Donut chart
		Pie chart
		Sankey diagram
		Stacked bar chart
		Tree map
Relationships	Show relationships and connections between the data or show correlations between two or more variables	Heat map
		Radar chart
		Venn diagram
	For showing connections	Network diagram
		Tree diagram
		Connection map
	For finding connections	Bubble chart
		Heat map
		Scatterplot
Hierarchy	Show how data or objects are ranked and ordered together	Tree diagram
		Tree map
		Sunburst diagram
Location	Show data across geographies	Bubble chart
		Choropleth map
		Connection map
		Dot map
		Flow map

(continued)

Table 10-4. (*continued*)

Part to a whole	Show parts of a variable to the total	Donut chart
		Pie chart
		Stacked bar chart
		Sunburst diagram
		Tree map
Distribution	Display frequency, how data is spread out over an interval or is grouped	Bubble chart
		Density plot
		Dot matrix chart
		Histogram
		Multi-set bar chart
		Pictogram chart
	Show geographic distribution	Dot map
		Connection map
		Flow map
	Show distribution by age and gender in a population	Population pyramid
	Show distribution in a body of text	Word cloud
How things work	Show how an object or system functions	Flow chart
		Illustration diagram
		Sankey diagram
Processes and methods	Explain processes or methods	Flow chart
		Gantt chart
		Illustration diagram
		Sankey diagram

(*continued*)

Table 10-4. (*continued*)

Movement or flow	Show the flow of data	Connection map Flow map Sankey diagram
Patterns	Reveal patterns in the data	Area chart Bar chart Bubble chart Candlestick chart Choropleth chart Connection map Dot map Heat map Histogram Line chart Multi-set bar chart Population pyramid Radar chart Scatterplot Stacked area chart
Range	Display variations between upper and lower limits on a scale	Bullet chart Candlestick chart Gantt chart Open high-low-close chart Span chart

(*continued*)

Table 10-4. *(continued)*

Data over time	Show data over a time period to determine trends or changes over time	Area chart Bubble chart Candlestick chart Gantt chart Heat map Line chart Open high-low-close chart Stacked area chart Stream graph
Analyzing text	Reveal patterns and insights from a body of text	Word cloud

Summary

This chapter provided a concise yet comprehensive guide to selecting the most suitable chart types for diverse visualization needs. It identified situations where different chart types excel and offered a mapping of chart categories, facilitating the identification of the ideal chart for specific purposes. The chapter also explained how to match the desired message with the appropriate visualization method by categorizing applications into comparisons, proportions, relationships, and patterns.

CHAPTER 11

Best Practices and Tips

Design

Provide an Effective Dashboard

The key objective is to provide an effective dashboard, not an artistic display that wows, which informs users with precisely what they need and in the way they need it. The needs and preferences of the users should be the top priority. Designing a dashboard layout is half art and half science. Be aware that for all types of dashboards, there will always be a battle between form and function.

Design a Single-Screen Dashboard

A dashboard should display its information on a single screen. From there, if needed, a user should be able to navigate to the detailed pages. A user should not have to click multiple times or navigate away from the main screen to secondary screens to view important information. When data is fragmented into separate screens, the benefit of displaying consolidated information at a glance, which enables one to see the whole picture, is lost. Not fitting all the important information on a single screen is a common mistake that should be avoided.

Design the Layout Carefully

A dashboard needs to be well designed from both a content and layout perspective. A common mistake is to arbitrarily cobble together information, which leads to a cluttered mess. A dashboard should be organized with appropriate information placement, based on the importance and desired viewing sequence, with a visual design that segregates data into meaningful groups.

A. Khan, *Visual Analytics for Dashboards*, https://doi.org/10.1007/979-8-8688-0119-8_11

The objective should not be limited to making the dashboard look good. The aim should also be to arrange the content so that it fits the way it is used. The most important data should be made prominent and items requiring immediate attention should stand out. Data to be compared must be arranged and visually designed in a way that encourages comparison.

Dashboard space is limited and, therefore, it is important that prime real estate is not wasted by leaving it blank or by using it for information that is not important.

Position Important Items Strategically

Without other visual cues, a user starts at the top left of the visual and then scans in a zig-zag motion across the page, as shown in Figure 11-1. Therefore, the design layout of individual items should take this into consideration.

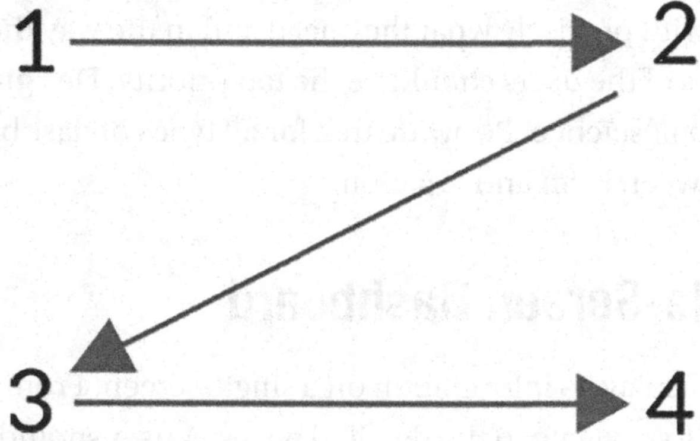

***Figure 11-1.** How information is scanned*

The top left of the page is prime real estate because that is the first area seen by the audience. Therefore, the most important item should be located there. The audience should not have to wade through other content to reach the important items.

Limit Display to What Is Essential

A dashboard should only display information that can be quickly scanned and understood. How much is too much depends a great deal on how it is presented. Some poorly designed displays cannot be scanned and understood in any amount of time, even though they contain only a few measures. If a chart is difficult to read, no reduction in the amount of information will change that fact. When properly designed, numerous values can be placed on a single chart in a way that is easy to scan and understand.

Start Design at the High Level

When designing the overall dashboard layout, start from a high level and then drill down to specific details, such as how to physically integrate data into the dashboard. The designer should understand how users think, their workflow, as well as how they view and operate their business. Users will not be satisfied with an out-of-the-box layout that fails to provide a level of visual customization suited to their business view.

Make the Dashboard Communicate Quickly

Users should be able to objectively assess the most important business measures with minimal interaction. If the data displayed is too granular or the screen is cluttered, the dashboard will be unable to quickly convey the state of affairs. The design should ensure that the dashboard is able to easily and quickly communicate with the users.

Pick Critical Metrics

Choosing metrics to include in the dashboard is a critical task, as they matter and should be relevant. Pick both leading and lagging indicators. Too many metrics make it difficult to analyze and take action. Therefore, the key metrics that are significant for the business should be identified and displayed on the dashboard, which will make it very valuable.

Enable Interactivity

A dashboard will place everyone who views it on the same page. Once there, viewers will have their own questions and areas about which they will want additional information. The dashboard should enable them to customize it so that they can obtain the needed information. Interactive, highly visual dashboards enable audiences, with little or no training, to perform basic analysis tasks such as filtering and drill-down for examining the underlying data. Viewers should be able to literally get the big picture from the dashboard and, subsequently, drill down into a view that provides them with what they need to get their job done.

Provide Intuitive Navigation

Consistency should be ensured, especially for the user interface and navigation. Attention should be paid to navigation between major UI items. Organize screens to ensure easy and intuitive navigation. Expect users to make mistakes. Therefore, the user interface should be designed to recover from user mistakes.

Make It Simple to Access and Use

Making dashboards easily accessible is critical. Users should be able to access a dashboard through different access methods and devices. These include web-based access, desktops, laptops, and mobile devices. The dashboard should be designed so that access through various methods does not lead to display or performance issues.

Keep It Current

The dashboard is frequently viewed as the single source of truth. Therefore, it should be ensured that the dashboard's underlying data is current, as outdated data can lead to a false sense of confidence. Users may believe they are making fact-based decisions, even though data may no longer be representative of or relevant to the current situation.

Develop on a Single Platform

It is easy for managers to build or buy their own solutions, independent of each other. Eventually, this leads to dashboard silos that compete for resources, especially from IT. To avoid the disruption of switching to a new platform, develop all dashboards and scorecards on a single platform, which leverages a unified data integration infrastructure. The platform should provide seamless support for the deployment of development, testing, and production environments. Also, the design should not be onerous for IT administration.

Information and Its Visual Representation

Avoid Mismatch

Information and its visual representation on a dashboard often become disconnected in two ways:

- Visual medium of representation is inappropriate.
- Visual representation of values does not match the values themselves.

For displaying information on a dashboard, the type of chart cannot be chosen randomly. Different types of charts are designed to display different types of information and to emphasize their unique features. Even an appropriate chart can end up being misleading. For example, a bar chart can provide the wrong perception if the quantitative scale does not start from zero. In such a case, only when the scale begins at zero will the bar heights accurately encode their relative values.

Display Context for the Data

Numbers by themselves are not very helpful for monitoring performance. Frequently, very little or no context is provided for the displayed data. If only numbers are provided, it can be difficult to understand them in isolation. Enough context should be provided so that it helps performance monitoring. Providing the appropriate context for the key measures makes the difference between numbers that are just displayed on the screen and those that inform and trigger action.

It is easy to show context on a dashboard, which can include goals, historical performance, as well as internal and external benchmarks. Express measures with enough context for their meaning to be clear, such as comparisons to targets, averages, and previous periods.

Limit Text or Content to Be Displayed

Context that will help the dashboard viewer determine whether or not action needs to be taken should be included. However, as much as possible, limit the amount of dashboard text or content that is truly static. Labels and text should be clear and indicate exactly what they mean from a business perspective.

Provide Qualitative Data

For monitoring the business, dashboards are usually predominantly populated with current quantitative measures. A mistake that many designers make is to focus exclusively on quantitative data while ignoring qualitative data. However, quantitative data should be supplemented with qualitative data, which provides a broader picture that can make the analysis more valuable.

Express Measures Effectively

A common mistake is that measures are not expressed in an effective way, as there is a lack of understanding regarding what the users need to see and how they plan to use the information. For a measure to be meaningful, users must know what is being measured and the units in which the measure is being expressed. A measure is poorly expressed if it fails to directly, clearly, and efficiently communicate the meaning to the dashboard user.

For example, a line chart can be developed, for budget versus actual revenue, as shown in Figure 11-2.

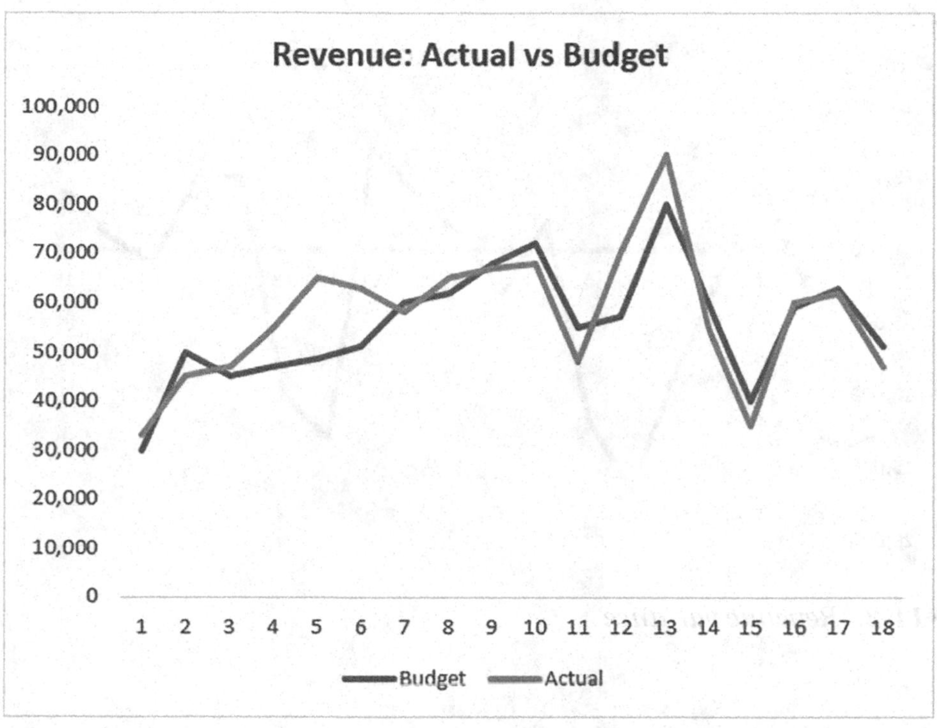

Figure 11-2. *Budget versus actual revenue*

This chart can be made more meaningful if the variance is expressed directly, as shown in Figure 11-3, which makes it easier for analysis.

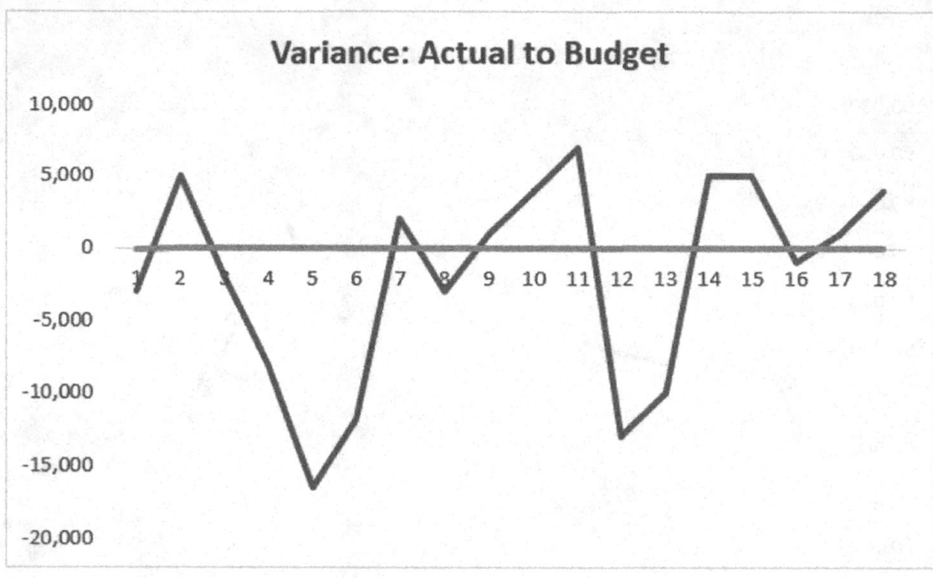

Figure 11-3. *Revenue variance*

Do Not Make Everything Prominent

All items displayed on a dashboard should be important, though they are not equally important. When a user views a dashboard, the eyes should immediately be drawn to the most important information, especially to the items that need immediate attention. Only the exceptions or items that need to be highlighted should stand out. If everything is visually prominent and vying for attention, nothing in particular will grab a user's attention, which will make the dashboard fall short of its objective.

Presentation Media and Variety

Unappealing Visual Display

A dashboard should be functionally effective and also be visually pleasing. However, many dashboards are unappealing or ugly, which puts their viewers in a frame of mind that is not positive for their effective use.

Using an Inappropriate Visual

This is a very common mistake on dashboards, as well as all forms of data presentation. When a designer picks the wrong type of chart such as a pie chart, when there are too many categories, it makes it useless from an analysis perspective. Using a chart when a table will work better or vice versa is a common mistake that should be avoided.

Unnecessary Display Variety

The benefit of display consistency is that it allows users to use the same strategy for interpreting data, which saves them valuable time and effort. Many designers display variety for the sake of variety, which is a common mistake. They believe that using the same type of visualization, such as bar charts, will be boring and project a lack of creativity. Display variety should not be encouraged without a good reason. The display type that works best should always be selected, even if that results in a dashboard filled with multiple instances of the same chart type.

Distractions That Hinder Communication

After deciding how the information and message are displayed, display components need to be designed so that communication is clear and efficient without any distractions. Among the design problems that can hinder communication are brightly colored bars and 3D effects, which make analysis difficult and challenging. Simple design mistakes like these can undermine a dashboard.

Data Issues

Excessive Detail or Precision

On a dashboard, every unnecessary piece of information wastes time and is unacceptable when time is of the essence. A dashboard should never display information more detailed or precise than necessary to support its objective of rapid monitoring. A user should never be forced to process data levels that are irrelevant to the task at hand. Too many details or measures expressed too precisely slow down users without providing any benefit.

As an example, instead of displaying a monthly revenue of $5,737,285.22, it can be represented as $5,737,285 because the two decimal points has no significance. An even better way will be to display the number as $5.73M, which helps avoid clutter.

Improper Encoding of Visual Objects

When a chart is used to communicate quantitative data, values are encoded in the form of visual objects, such as bars. Visual objects should accurately encode the values so that they can be compared to one another and enable the understanding of relationships. Sometimes graphical representations of quantitative data are poorly designed, in ways that display quantities inaccurately or distort the data, such as using a bar chart y-scale that does not start from zero.

Visualization Tips

Keep It Visual

Dashboards are provided for quick analysis, as report and text-based tables are not fast or easy to read. When developing a dashboard, remember the adage, "A picture is worth a thousand words." The human brain processes a number, visualization, or picture as individual chunks of information. Therefore, a report or table filled with numbers requires the brain to store and remember multiple chunks. However, a visualization or picture requires a single chunk.

The process of comprehension and insight is dramatically faster with visualizations. If they are provided, users can actually focus on what the dashboard views are conveying, instead of trying to determine how to read and interpret views. A well-designed, highly visual dashboard is more widely adopted by audiences.

Leverage Visual Perception

When designing a dashboard, include colors, shapes, lines, thicknesses, shading, and any other tools that leverage visual perception. Avoid overly cute widgets, 3D graphics, and chart types not commonly used. Visualizations like bar charts, line charts, heat maps, and scatterplots are popular because they are clear and easy to understand, and everyone knows how to interpret them.

Avoid Distracting Visuals

Dashboards are displays used to monitor important information, critical at times, and not an appropriate venue for artistic expression. Therefore, dashboard visuals should not overwhelm or distract. Dashboard information should stand out clearly without competition. Partly due to their visual nature, people have a tendency to dress up dashboards with all sorts of visuals that do not contribute to any analysis. Any visual content that does not express data or is unnecessary for supporting data presentation in some useful way is a distraction.

Avoid Inappropriate Visual Salience

Visual salience is the distinct subjective perceptual quality that makes some objects stand out from their adjacent items and, therefore, grab the viewer's attention. If everything is eye-catching, the result is that nothing stands out. Therefore, inappropriate visual salience should be avoided. Control the visual salience of the information to support each item's relative importance.

All information that deserves space on a dashboard is important or should be. However, not all information is equally important. Also, items not important on a routine basis could become extremely important when something goes wrong and, therefore, demand attention. A designer should incorporate ways of increasing the visual salience of items when conditions demand that they stand out.

Mute Axis Lines, Borders, and Lines

The axis lines used to define the data region of a chart can be quite useful. However, they can be muted without any negative effect, as shown in Figure 11-4.

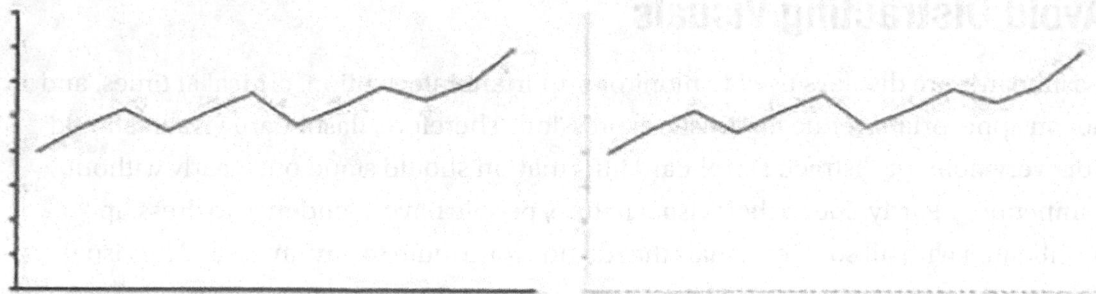

Figure 11-4. *Muting axis lines*

Lines, borders, or fill colors can be used to delineate data sections when white space is insufficient. In such a case, keep line weight and color intensity to a minimum, as shown in the lower section of Figure 11-5.

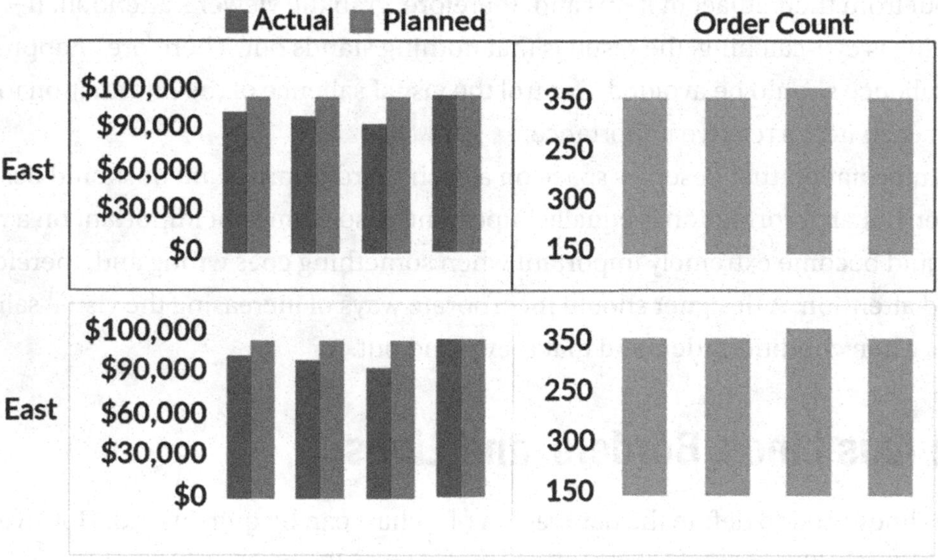

Figure 11-5. *Muting line borders*

Avoid Grid Lines

Grid lines, which are rarely useful, are among the most prevalent form of distracting non-data pixels found on dashboards. Grid lines in tables, especially thick ones, can make displays visually difficult. Use grid lines and/or fill color in tables when white space alone cannot adequately delineate columns and/or rows. In Figure 11-6, the table on the right should be favored compared to the one on the left.

	Jan	Feb	Mar	Apr
Product 1	64,377	93,732	404,762	24,871
Product 2	324,100	7,513	37,840	343,984
Product 3	36,786	-21,790	59,322	91,607
Product 4	431,749	95,053	50,792	163,344
Product 5	387,043	101,230	49,312	173,654
Total	1,244,055	275,738	602,028	797,460

	Jan	Feb	Mar	Apr
Product 1	64,377	93,732	404,762	24,871
Product 2	324,100	7,513	37,840	343,984
Product 3	36,786	-21,790	59,322	91,607
Product 4	431,749	95,053	50,792	163,344
Product 5	387,043	101,230	49,312	173,654
Total	1,244,055	275,738	602,028	797,460

Figure 11-6. *Grid lines in tables*

Use Fill Colors to Delineate

Use fill colors in alternate rows to delineate rows when the rows are wide, as shown in Figure 11-7, as it makes tracking easier.

	Jan	Feb	Mar	Apr	May	Jun
Product 1	64,377	93,732	404,762	24,625	42,293	647,828
Product 2	324,100	7,513	37,840	343,984	353,733	198,494
Product 3	36,786	-21,790	59,322	91,607	24,811	401,451
Product 4	431,749	95,053	50,792	163,344	309,731	198,494
Product 5	387,043	101,230	49,312	173,654	427,726	120,213

Figure 11-7. *Highlighting table rows*

Figure 11-8 shows a table where grid lines and fill shading are used to delineate columns and rows. The lower version, which is muted, should be favored.

Product	Jan	Feb	Mar	Q1	Apr	May	Jun	Q2	YTD
Product 1	64,377	93,732	404,762	562,871	24,871	42,293	647,828	714,992	1,277,863
Product 2	324,100	7,513	37,840	369,453	343,984	353,733	198,494	896,211	1,265,664
Product 3	36,786	-21,790	59,322	74,318	91,607	24,811	401,451	517,869	592,187
Product 4	431,749	95,053	50,792	577,594	163,344	309,731	198,494	671,569	1,249,163
Product 5	387,043	101,230	49,312	537,585	173,654	427,726	120,213	721,593	1,259,178
Total	1,244,055	275,738	602,028	2,121,821	797,460	1,158,294	1,566,480	3,522,234	5,644,055

Product	Jan	Feb	Mar	Q1	Apr	May	Jun	Q2	YTD
Product 1	64,377	93,732	404,762	562,871	24,871	42,293	647,828	714,992	1,277,863
Product 2	324,100	7,513	37,840	369,453	343,984	353,733	198,494	896,211	1,265,664
Product 3	36,786	-21,790	59,322	74,318	91,607	24,811	401,451	517,869	592,187
Product 4	431,749	95,053	50,792	577,594	163,344	309,731	198,494	671,569	1,249,163
Product 5	387,043	101,230	49,312	537,585	173,654	427,726	120,213	721,593	1,259,178
Total	1,244,055	275,738	602,028	2,121,821	797,460	1,158,294	1,566,480	3,522,234	5,644,055

Figure 11-8. *Muted table display*

Charts

Use the most appropriate type of chart for the data being analyzed. Give charts meaningful titles and make them more readable. This can be done by reducing non-data pixels and, where possible, eliminating all unnecessary data and non-data pixels.

Gauges

For gauge pointers, be careful with using similar colors. Use a legend to distinguish pointers. If multiple gauges show the same metrics, use a global legend. Use a neutral color for the gauge background. Pick a bold color for the actual value pointer and a complementary color for the target. Use colors that depict what the range implies, such as green for good.

Clutter and Color Mistakes

Avoid Screen Clutter

Clutter refers to visual elements that take up space but don't increase understanding. Most dashboards are cluttered with unnecessary content and decoration. They make it difficult to quickly find the needed information or perform analysis.

The presence of clutter in visual communications can cause a less-than-ideal or uncomfortable user experience. It can make something feel more complicated than it actually is. When a visual feels complicated, there exists the risk that the audience may decide that it does not want to spend time trying to understand what is being displayed. In such a case, the ability to communicate with the audience is lost.

Due to their visual nature, even though visual distractions may appeal initially, users soon view them as a hindrance to their ability to quickly perform their work. Users must not need to process to get to the data. Therefore, carefully review the visual elements contained in the dashboard. Identify items that do not add information value and remove them.

Avoid Misusing or Overusing Color

Color can be used in powerful ways to highlight data, encode data, or create a relationship between individual dashboard items. However, it is frequently overused and misused. An example of color overuse is using different colors for bars when only one category is involved. In such a case, the bar height is the best indicator of value and color provides no value. Using color actually slows the typical user who tries to determine its significance where none exists.

Using too many colors together causes their pre-attentive value to be lost. When used sparingly, color is one of the most important tools for drawing attention. Resist the urge to use color for the sake of being colorful. The use of too many colors can be visually overwhelming, and when overused, color loses its ability to highlight what is most important. Leverage color, which should be used consistently and selectively as a strategic tool to highlight important parts of the visual. When color is used, follow the contrast rule and ensure that the screens are still readable.

Avoid Unnecessary Color Variation

Use differences in color only to indicate differences in data and only when some other visual means will not work as well. For example, if the dashboard information naturally falls into four different groups, use a different color in the background of each section to delineate them. Alternatively, light borders or perhaps even white space alone will do the job without an unnecessary abundance of color.

Select Colors Carefully

The colors selected should be based on an understanding of how users perceive color. Some colors are hot and demand attention, while others are cooler and less visible. When any color appears as a contrast to the norm, the eyes pay attention and the brain attempts to assign meaning to that difference. When selecting colors, keep in mind that approximately 8% of males and 1% of females are color blind or deficient.

Since dashboards are often densely packed with information, visual content must be kept as simple as possible. Colors should be subdued and neutral unless they are being used to highlight something important. If color is used sparingly, and reliance is mostly on soft, natural, and relatively neutral colors such as gray for most items, then the stage is set for using bright and dark colors to draw attention when needed.

In general, start with shades of gray and pick a single bold color to draw attention where it is needed. Start with a base color of gray, not black, which allows for greater contrast since color stands out more against gray than black. Blue is a good attention-grabbing color as it avoids the colorblindness issue and also prints well in black and white.

Maintain Color Consistency

It does not make sense to change colors to prevent boredom or provide variety for the sake of variety. Use color changes only to signal change. If color consistency is maintained, the users will take time to familiarize themselves with what the color means. After that, they will assume that the same details apply throughout the communication.

Visuals to Avoid

Pie Charts

Pie charts should be avoided as it is difficult to accurately interpret data with this type of visualization. The human eye is not good at ascribing quantitative value to two-dimensional space. In other words, pie charts are difficult to analyze. When segments are similar in size, it is difficult to determine which slice is bigger. In some cases, it is possible to determine which one is larger but not by how much.

An alternative solution is to use a bar chart where the eye compares the end points. Since they are aligned to a common baseline, it is easy to assess their relative value. That makes it straightforward to determine not only which segment is largest but, also, how incrementally larger it is than the other segments.

Donut Charts

Avoid donut charts like pie charts. With this type of chart, the user has to compare one arc length to another arc length, like arc A to arc B in Figure 11-9, compared to angles and areas for pies.

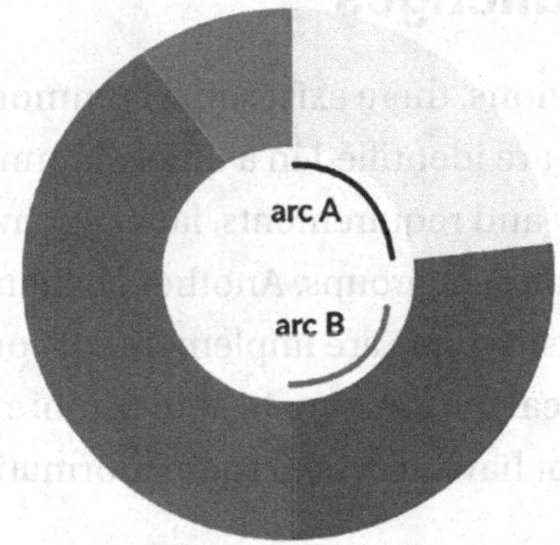

***Figure 11-9.** Donut chart arcs*

3D Charts

Never use 3D charts, especially to plot a single dimension. The only exception is when a third dimension needs to be plotted. However, even then, it can get tricky really quickly. Like the pie chart, 3D skews the numbers, as it makes them difficult or impossible to interpret or compare. Adding 3D to graphics introduces unnecessary chart elements like side and floor panels.

Charts with Secondary Y-Axis

With this type of chart, it takes time and effort to determine which data should be read against which axis. Therefore, avoid the use of a secondary or right-hand side y-axis.

Challenges

Unique Business Operations

Dashboards, which are often unfamiliar to many business users, cater to the needs of a wide variety of users ranging from the C-suite to lower-level employees in the organization's hierarchy. Also, businesses operate differently and, hence, their dashboards will reflect that reality. Therefore, dashboards must be iteratively customized to fit different individual business needs.

Organizational Challenges

For dashboard implementations, there exist some common challenges, which are not difficult to overcome if they are identified in a timely manner. These include budget limitations, unclear strategy and requirements, lack of standardization, and poor communication between different groups. Another challenge is political, as boundaries can get eliminated when dashboards are implemented. Some managers feel a loss of control when business-critical information becomes more widely available, especially to those who previously did not have access to such information.

User Acceptance

In many cases, dashboards may not be the only solution for analyzing and acting upon key business information. Users may feel more comfortable with legacy systems and spreadsheets. They may not feel that dashboards are essential since there are other ways of getting the job done, even if the older methods are actually more time-consuming and inefficient. Therefore, when implementing dashboards, change management should be used to achieve user acceptance.

Data Quality

The importance of data quality cannot be over-emphasized. The reality is that data quality issues are constantly underestimated. Usually, it is much worse than anyone believes. When poor data quality leads to project development issues or incorrect results post-implementation, users lose confidence in the dashboard. Without user acceptance, no dashboard can succeed.

Speedy Implementations

Some managers don't want to involve IT and initiate a well-planned project, which takes time and is costly. They just want something quick, simple, and cheap to replace their Excel spreadsheet or numerous reports, which they must comb through to perform any analysis. Such dashboards can be quite attractive visually and potentially powerful. However, there is usually a long-term tradeoff for short-term gain. These dashboards do not scale and frequently have to deal with data quality issues.

Summary

This chapter described the best practices for dashboards pertaining to design, information and its visual representation, presentation, and data. It also provided useful visualization tips, such as leveraging visual perception, avoiding distracting visuals, and avoiding inappropriate visual salience. The chapter also discussed common mistakes pertaining to clutter and color misuse and overuse and identified visuals that should be avoided. Additionally, the chapter addressed challenges such as organizational adaptation, user acceptance, and data quality.

Appendix

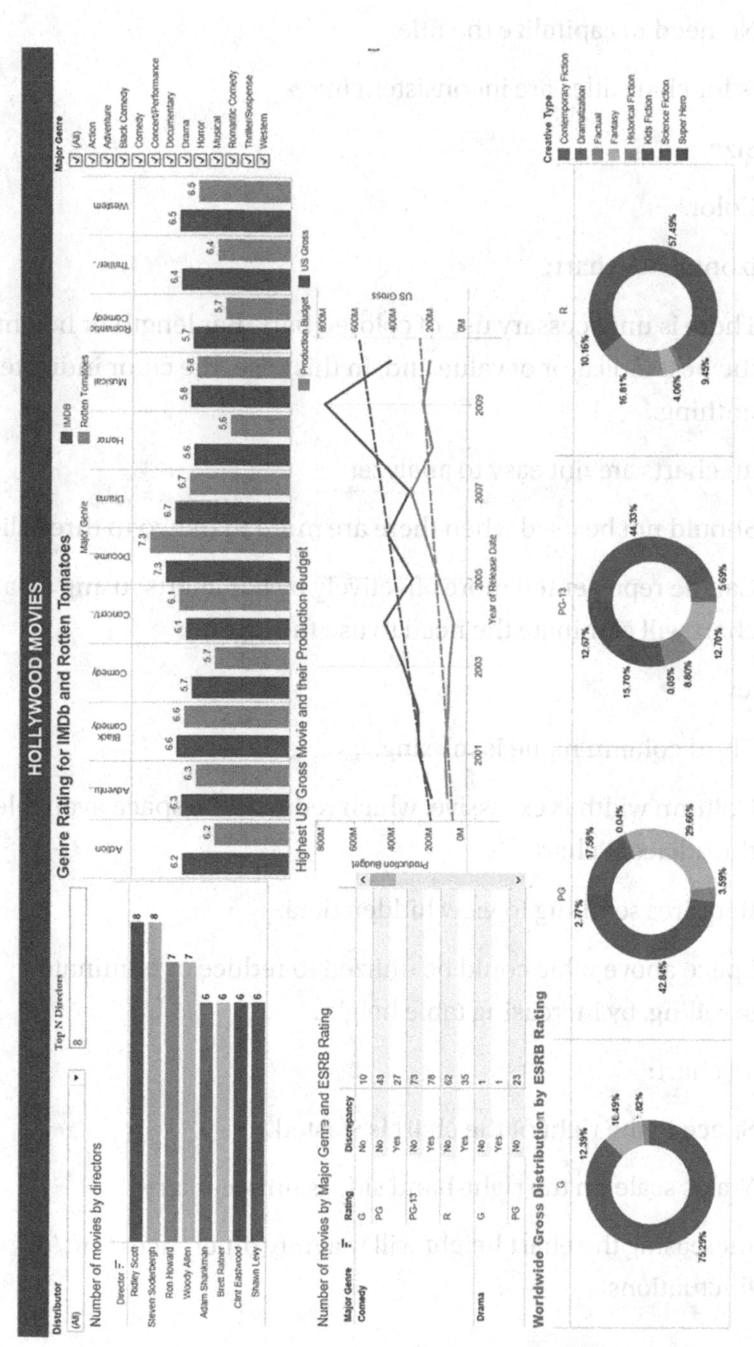

A. Khan, *Visual Analytics for Dashboards*, https://doi.org/10.1007/979-8-8688-0119-8

Dissecting the Dashboard

This dashboard has many issues, which are listed here:

- Dashboard title:
    - No need to capitalize the title.
- Fonts for chart titles are inconsistent in:
    - Size.
    - Color.
- Horizontal bar chart:
    - There is unnecessary use of colored bars. Bar length or height is the best indicator of value and, in this case, the color indicates nothing.
- Donut charts are not easy to analyze:
    - Should not be used when there are more than two to three slices.
    - Can be represented more effectively by bar charts; using a bar chart will eliminate the need to use the legend.
- Table:
    - Third column name is missing.
    - Column width is excessive, which reduces the space available for the adjacent chart.
    - Requires scrolling to view hidden data.
    - Space above table could be utilized to reduce or eliminate scrolling, by increasing table height.
- Trend chart:
    - Space to the right of the chart is wasted.
    - Y-axis scale on the right-hand side is unnecessary.
    - Increasing the chart height will magnify differences and/or fluctuations.

- Legend:
    - IMDB (blue) and Rotten Tomatoes (red) can be placed side by side in the vertical bar chart (just like in the trend chart).
- Filters (Distributor and Top N Directors):
    - Excessive width, which can be reduced.
- Clutter:
    - No need to place numbers at the top of each bar.
- Vertical bar chart:
    - Y-axis scale is missing.
- Top left-hand corner of the screen:
    - Shows "Number of movies by Director." If this is the most important visualization, then it is correctly placed. However, if the trend chart is the most important item, then it should be placed in the top left-hand corner of the screen.

Index

A. Khan, *Visual Analytics for Dashboards*, https://doi.org/10.1007/979-8-8688-0119-8

C

D

E

F

T

U

V

W, X, Y

Z

Printed in the United States
by Baker & Taylor Publisher Services

Printed in the United States
by Baker & Taylor Publisher Services